ATOMIC GREEN

Nuclear Power Can Stop Climate Change

BY MARY FRAN REED, PhD

To everyone who cares enough about our future and the future of our planet to do something about it.

Table of Contents

Chapter 1

Introduction: Climate Change is Already Here!

I'll start this book with what I consider to be the major problem that we, as the citizens of the world, are currently facing. Here are just a few of the many recent news headlines highlighting the effects of global warming and the extreme weather events we are currently experiencing.

The Current Effects of Global Warming

- "Tornado Updates: At Least 32 Dead, Dozens Injured Across 9 States" - ABC News, April 14, 2023. Tornadoes caused significant casualties and damage across nine states.

- "U.N. Climate Scientists Conclude this is the 'Hottest Period in the Last 125,000 Years'" - CBS Face the Nation, June 4, 2023. U.N. climate scientists conclude that this is the hottest period in the last 125,000 years.

- "Melting Arctic Ice Threatens Polar Bear Populations: Climate Change Impacts Evident in the North" - National Geographic, July 30, 2023. The Arctic is experiencing rapid ice melt, threatening polar bear populations and other wildlife.

- "Floods Devastate South Asia: Climate Change Amplifies Monsoon Rains, Affecting Millions" - Al Jazeera, August 1, 2023. Climate change has amplified monsoon rains in South Asia, causing devastating floods.

- "Unprecedented Wildfires Ravage California: Experts Warn of Worsening Fire Seasons Due to Climate Change" - The Guardian, August 10, 2023. Wildfires in California reached unprecedented levels, with experts warning of worsening conditions due to climate change.

- "Coral Reefs in Crisis: Global Warming Triggers Mass Bleaching Events and Loss of Marine Biodiversity" - Science Daily, August 25, 2023. Global warming has triggered mass bleaching events and the loss of marine biodiversity in coral reefs.

- "Climate Change Impacting Agriculture: Erratic Weather Patterns Cause Crop Failures and Food Price Volatility" - Bloomberg, September 2, 2023. Erratic weather patterns due to climate change are causing crop failures and food price volatility.

- "Hurricane Season Intensifies: Powerful Storms Hit Atlantic as Climate Change Fuels More Destructive Weather" - CNN, September 5, 2023. The Atlantic hurricane season saw powerful storms, intensified by climate change.

- "Floods, Fires, and Deadly Heat are the Alarm Bells of a Planet on the Brink" - The Washington Post, September 27, 2023. The increasing frequency and intensity of floods and fires signal the urgent need for climate action.

- "Extreme Droughts Grip Parts of Africa: Climate Change Linked to Water Scarcity and Food Insecurity" - Reuters, October 12, 2023. Parts of Africa are experiencing extreme droughts, linked to climate change, causing water scarcity and food insecurity.

- "Earth Facing Dire Sea Level Rise — Up to 20m — Even if Climate Goals are Met" - Global News, November 17, 2023. New modeling data from the International Cryosphere Climate Initiative indicates sea level rises of up to 20 meters even if the global temperature rise is limited to 2 degrees Celsius.

- "2023 Was the World's Warmest Year on Record, by Far" - NOAA, December 2023. 2023 surpassed previous records, becoming the warmest year globally.

- "Global Risks Report 2024: Climate-Related Threats Dominate" - World Economic Forum, January 2024. Climate change remains one of the top risks facing global populations, with extreme weather events expected to increase.

- "Marine Heatwave in the Pacific Causes Mass Coral Bleaching" - Science Daily, February 2024. A marine heatwave has led to widespread coral bleaching in the Pacific, threatening marine biodiversity.

- "Record-Breaking Heat Hits Western Australia, Causing Extreme Fire Danger" - Sydney Morning Herald, February 23, 2024. Western Australia experienced its hottest night on record, leading to extreme fire danger and school closures.

- "Severe Flooding and Droughts Expected to Worsen Due to Climate Change" - Carbon Brief, February 23, 2024. Regions from Asia to Africa are experiencing extreme weather patterns exacerbated by climate change.

- "Rising Sea Levels Threaten Coastal Communities Worldwide" - The New York Times, April 2024. Coastal flooding is becoming more frequent and severe due to rising sea levels.

- "Climate Change-Driven Floods and Fires are Alarm Bells for a Planet on the Brink" - The Washington Post, April 2024. The increasing frequency and intensity of floods and fires signal the urgent need for climate action.

- "Asia Hit Hardest by Climate Change and Extreme Weather" - UN News, April 23, 2024. Extreme conditions from droughts and heatwaves to floods and storms have severely impacted Asia, with 2023 marking the hottest year on record for many countries in the region.

- "Deadly Floods in UAE and Oman Linked to Climate Change" - Al Jazeera, April 25, 2024. Intense rains caused by global warming resulted in fatal floods in the UAE and Oman, with the heaviest rainfall in 75 years.

- "El Niño-Induced Extreme Weather Devastates Southern Africa" - UN News, April 29, 2024. The El Niño event has caused severe flooding and drought, affecting millions across southern Africa.

- "Extreme Weather Events Causing Food Insecurity for Millions" - UN News, April 29, 2024. Climate-related extreme weather is one of the main drivers of food insecurity for 72 million people in 18 countries.

- "Heat Exposure of Older People Across the World to Double by 2050, Finds Study" – Guardian, May 14, 2024. An extra 270 million adults aged 69 or over will suffer dangerous heat levels of 37.5°C amid global heating and aging populations.

- "Passenger Dead, 30 Injured as Singapore Airlines Flight Hits Severe Turbulence" - CBC News, May 21, 2024. A Singapore Airlines flight from London to Singapore encountered sudden and severe turbulence, resulting in the death of a 73-year-old British man and injuries to 30 passengers.

- "Heatwaves Cause Public Health Crisis Across Europe" - BBC News, June 2024. Europe faces severe health challenges as record-breaking heatwaves continue to strike.

- "Record-Breaking Global Warming: The World Heads Towards 'Climate Hell' as Temperatures Soar" by Kate Abnett and Gabrielle Tétrault-Farber – Reuters, June 5, 2024. Each of the past twelve months has been the warmest on record in year-on-year comparisons, according to the European Union's Copernicus Climate Change Service. U.N. Secretary-General António Guterres emphasized the urgency of the situation, calling for immediate action to prevent what he described as "climate hell."

- "The oceans are heating so fast, some scientists call for a new "Category 6" hurricane classification" – Bulletin of the Atomic Scientists, July 11, 2024. Hurricane Beryl, which slammed into Texas after wreaking havoc in the Caribbean, was supercharged by "absolutely crazy" ocean temperatures that are likely to fuel further violent storms in the coming months, scientists warned.

- "Monday breaks the record for the hottest day ever on Earth" – Associated Press, July 24, 2024. Driven by oceans that won't cool down and worsening climate change humans are experiencing the hottest days ever measured.

- "On one night, two places in the Northeast get hit with 1-in 1,000 year rainfall" – NBC News, August 19, 2024. Climate change has made more severe storms more likely in the Northeast because a warmer atmosphere can hold more water.

- "Hurricane Helene and the 'Known Unknown' of Climate Costs" –Time, October 4, 2024. Hurricane Helene is a significant human tragedy with more than 200 deaths reported as of Oct. 4. Thousands more have been left homeless. And the property damage will be difficult for the thousands of homeowners to manage without adequate flood insurance. Estimates of the economic damage vary. Moody's initially put the price tag as high as $34 billion. Others, which take into account lost economic output, say the toll could top $100 billion.

Every one of these headlines is a stark reminder of global warming's impact: rising seas, melting ice, droughts, wildfires, devastating storms, and unprecedented heat waves disrupting ecosystems and human life. Each day brings more evidence that the climate crisis is accelerating, and unless we drastically reduce greenhouse gas emissions, it will spiral into an even graver catastrophe.

In 2007, Al Gore's book, *An Inconvenient Truth,* issued a powerful warning, vividly depicting the melting polar ice, rising sea levels, and intensifying natural disasters. His message was clear: immediate action was necessary to avert catastrophe. His book served as a wake-up call, exposing the threats to human health, the environment, and global economies if we stayed on our current path.

Since then, progress has been inconsistent. While some have resisted change, fearing the economic costs of transitioning to cleaner energy, others have fought for transformative climate policies. Despite resistance, we've seen meaningful advances in legislation and international agreements. But the question remains: will we act boldly and swiftly enough to prevent disaster? The clock is ticking, and the future of our planet hangs in the balance.

The Paris Agreement

The Paris Agreement, adopted in 2015, is a legally binding international treaty with a singular mission: combat climate change by limiting global warming to "well below" 2°C (3.6°F), with a clear emphasis on keeping it as close as possible to 1.5°C (2.7°F) compared to pre-industrial levels. Key commitments under the agreement include setting ambitious temperature goals, requiring countries to establish their own emissions reduction targets—known as Nationally Determined Contributions—which are reviewed every five years, and striving to achieve global net-zero greenhouse gas emissions in the latter half of this century.

Recent developments underscore the urgency of action. The first global "taking stock" in 2023 revealed that the world is dangerously off track in meeting the Paris Agreement's long-term goals. Initiatives like the Global Methane Pledge, which aims to cut methane emissions by 30% from 2020 levels by 2030, represent significant steps forward, particularly with strong commitments from the oil and gas industry.

Scientists have repeatedly emphasized the critical importance of the 1.5°C target. Even slight temperature increases pose significant threats: prolonged heatwaves, intensified storms, wildfires, and other extreme events. The 1.5°C target is not just a number—it is a threshold beyond which the impacts of global warming could very well become catastrophic and irreversible.

International cooperation and vigilant monitoring are the lifeblood of the Paris Agreement. Countries convene annually to assess progress and ramp up their climate commitments. A robust system for tracking national commitments ensures transparency and accountability, making sure that all efforts are documented and reviewed. Overall, the Paris Agreement represents a global effort to unite nations in the fight against climate change, with a growing ambition to meet these critical targets.

However, significant challenges arose when former President Donald Trump announced the U.S. withdrawal from the Paris Agreement on June 1, 2017. This decision was met with widespread criticism and dismay from environmentalists, scientists, and other nations committed to combating climate change. Trump argued that the agreement unfairly disadvantaged the United States economically and was a bad deal for American workers.

The withdrawal process, as outlined by the terms of the Paris Agreement, became effective on November 4, 2020, just one day after the U.S. presidential election. During the intervening period, the United States remained a party to the agreement, albeit with limited participation.

Upon assuming office on January 20, 2021, President Joe Biden acted swiftly, signing an executive order to rejoin the Paris Agreement. This move was part of his broader agenda to prioritize climate action and reestablish the United States as a global leader in the fight against climate change.

The delay caused by the U.S. withdrawal under the Trump administration had several adverse effects. Firstly, it significantly undermined international efforts to address climate change. As the world's second-largest emitter of greenhouse gases, the United States' absence weakened the overall global response to the climate crisis.

Secondly, the delay hampered progress toward implementing the goals outlined in the Paris Agreement. This agreement relies on collaborative efforts among countries to reduce greenhouse gas emissions, enhance adaptation measures, and provide financial support to developing nations. Without the active participation of the United States, which possesses significant technological, financial, and political resources, achieving these targets became increasingly difficult.

Lastly, the withdrawal tarnished the perception of the U.S. commitment to addressing climate change. The move was seen by many as a retreat from the U.S.'s global leadership role in climate action, with potential diplomatic and reputational consequences. Other countries may have become less inclined to cooperate on climate-related issues or to follow the U.S.'s lead in implementing climate policies.

However, with President Biden's reentry into the Paris Agreement, the U.S. signaled a renewed and urgent commitment to climate action. This decision was widely welcomed by the international community and has helped restore confidence in global climate efforts. The Biden administration has since set ambitious climate goals, including achieving net-zero greenhouse gas emissions by 2050 and implementing robust domestic policies to support these targets.

It's important to highlight that while challenges remain, there has been substantial progress globally in the adoption of renewable energy sources and the reduction of carbon emissions. The rapid growth of solar and wind power industries, advancements in battery technology, and increased public awareness have contributed to positive momentum. Additionally, numerous countries, cities, and states have taken the lead in implementing ambitious climate action plans and setting stringent emission reduction targets.

In summary, the years since the Paris Agreement was adopted have seen a growing recognition of the need to address global warming, leading to a shift in public opinion and increasing pressure on policymakers to take concrete action. The scientific consensus on climate change has strengthened, and public awareness has significantly increased, driving these efforts forward.

On August 16, 2022, President Biden signed the Inflation Reduction Act—a landmark piece of legislation in our fight against climate change. This law includes $369 billion in investments for climate and clean energy initiatives, making it the largest and most ambitious climate legislation ever passed by Congress. This monumental investment provides a glimmer of hope for our future.

Understanding the Urgency of Global Warming

The world is confronting a massive crisis called global warming, and the stakes couldn't be higher. Rapid changes in our climate are becoming the new normal, and the reality is, we, humans, are the main culprits. Our relentless burning of fossil fuels, widespread deforestation, and industrial activities have dangerously increased the levels of greenhouse gases—like carbon dioxide and methane—

in the atmosphere, trapping heat from the sun and pushing global temperatures to unprecedented levels.

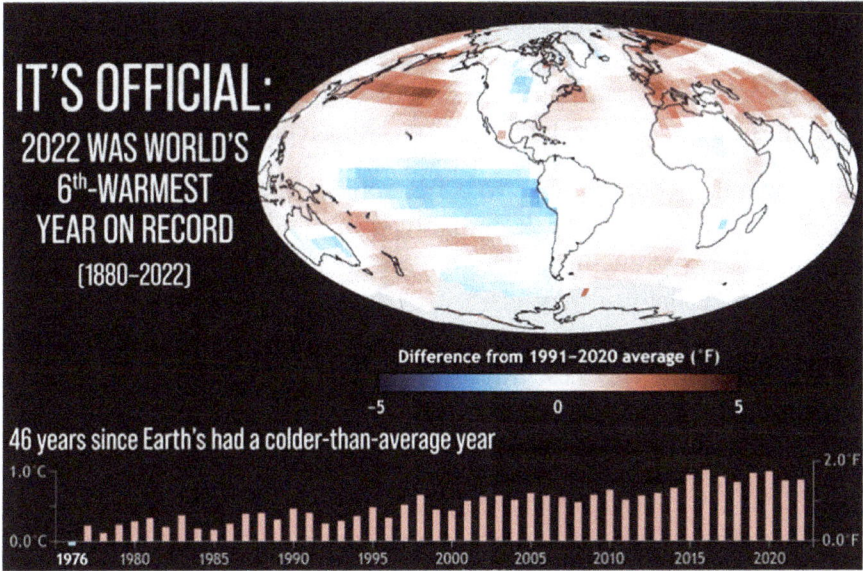

The bars on this graph illustrate the rising global temperatures compared to the 20th-century average, spanning from 1976 (left), the last year the world experienced cooler-than-average temperatures to 2022 (right). Data from NOAA's National Centers for Environmental Information starkly highlights the trend: our planet is heating up at an alarming rate. https://www.climate.gov/media/15007

Net-Zero Emissions

Achieving net-zero emissions is no longer a lofty ideal—it is an absolute necessity driven by the urgent need to combat climate change and its far-reaching consequences. At its core, net-zero aims to balance the amount of greenhouse gases emitted with an equivalent amount removed from the atmosphere, effectively halting the buildup of these heat-trapping gases that fuel global warming. The primary goal? To limit the rise in global temperatures to 1.5°C above pre-industrial levels—a threshold beyond which scientists warn of severe and potentially irreversible impacts on our planet's ecosystems and human societies.

But the push for net-zero emissions isn't just about stabilizing the climate; it offers a host of environmental and health benefits. By reducing pollution, particularly in urban areas, we can significantly improve air quality, leading to major public health gains, including a reduction in respiratory and cardiovascular diseases. Moreover, lowering carbon dioxide emissions helps mitigate ocean acidification, safeguarding marine ecosystems that are vital for global biodiversity and food security.

Economically, transitioning to a net-zero future presents both challenges and opportunities. It demands a fundamental shift in how we produce energy, how industries operate, and how consumers behave. While this transition may disrupt traditional industries, it also fuels innovation and creates new job opportunities in green technologies and sustainable practices. Importantly, achieving net-zero emissions must be pursued with a commitment to social equity, ensuring a just transition that supports vulnerable communities and workers in high-carbon industries.

The global momentum for net-zero emissions is growing. Many countries, cities, and businesses are setting ambitious targets aligned with international agreements like the Paris Agreement. This collective effort underscores a crucial recognition: addressing climate change requires coordinated action across all sectors of society. Ultimately, the pursuit of net-zero emissions is a critical investment in our planet's future, aiming to secure a sustainable and livable world for current and future generations.

Conclusion: The Time to Act is Now

Let's not sugarcoat it—global warming is no joke. As the headlines scream, we're already witnessing glaciers melting, sea levels rising, extreme weather events becoming more frequent and intense, and dramatic shifts in ecosystems and biodiversity. The above graph represents up to 2022, but 2023 was hotter yet and 2024 is turning out to be even hotter, with more all-time records being shattered over and over again. Global warming is already affecting our planet and all its inhabitants, including us.

I cannot stress enough how critical it is for us to act immediately. If we delay, the situation will only worsen, and the consequences could become irreversible. Scientists are in consensus: we need immediate and significant reductions in greenhouse gas emissions to keep global warming within manageable levels.

This brings us to the urgent need for mitigation. We must reduce those greenhouse gas emissions and minimize the extent of global warming. How do we achieve this? It involves transitioning to cleaner and more sustainable energy sources, improving energy efficiency, and implementing policies that limit carbon emissions across various sectors. We need to maximize our use of solar, wind, hydroelectric, and other low-carbon energy sources.

Here's where nuclear energy enters the conversation. In fact, nuclear energy isn't just part of the conversation—it's the most critical tool we have in the fight to stop climate change. In a world that's on the brink, with fossil fuels pushing us further toward environmental catastrophe daily, nuclear power stands apart. Why? Because it produces extremely low levels of carbon emissions, and during electricity generation, it releases no carbon dioxide at all. This isn't just an advantage—it's a necessity if we're serious about decarbonizing our economy and averting climate disaster.

Nuclear energy is not just impressive—it's indispensable. Throughout this book, I will show you how this powerful technology can be our best weapon in the battle to preserve our planet. We no longer have the luxury of time to debate incremental changes; we need solutions that work *now*, and nuclear energy is that solution.

In this book, written for those who care deeply about the future of our planet, I'll reveal why nuclear energy must play a pivotal role in mitigating climate change. We'll dive into the causes and consequences of global warming, analyze the current status of green energy efforts, and explore the urgency for bold, decisive action. Above all, we'll focus on how nuclear energy can—no, *will*—be the key to a carbon-free future.

This isn't a complex or abstract debate. It's a straightforward reality: nuclear power is not just part of the solution; it is *the* solution that can meet the scale and urgency of this global crisis.

Understanding the drivers of global warming, such as greenhouse gas emissions and their effects on weather patterns, is critical. In the next chapter, we delve into the specifics of how these emissions impact our planet, highlighting the critical role of natural climate phenomena like El Niño and La Niña in exacerbating these challenges.

References

1. *The Paris Agreement | UNFCCC.* (n.d.). https://unfccc.int/process-and-meetings/the-paris-agreement
2. NCEI.Monitoring.Info@noaa.gov. (n.d.). *Climate at a Glance | Global Mapping | National Centers for Environmental Information (NCEI).* https://www.ncei.noaa.gov/access/monitoring/climate-at-a-glance/global/mapping
3. *About NOAA Climate.gov.* (n.d.). NOAA Climate.gov. https://www.climate.gov/about
4. *Is it possible to achieve net-zero emissions?* (2021, October 7). National Academies. https://www.nationalacademies. org/based-on-science/is-it-possible-to-achieve-net-zero-emissions

Chapter 2

The Current Problem: Exploring the Impact of Greenhouse Gas Emissions

Greenhouse gas emissions are the invisible force reshaping our world. While their role in global warming is widely recognized, their interplay with natural climate phenomena like El Niño and La Niña presents even greater challenges. These patterns disrupt global weather systems, intensifying droughts, floods, and wildfires. Here, we explore these dynamics in detail and their cascading effects on agriculture, ecosystems, and human livelihoods. By understanding the depth of this challenge, we can better appreciate the essential and urgent need for a comprehensive range of solutions, including nuclear power.

Greenhouse Gases

Greenhouse gases (GHGs) play a crucial role in regulating the Earth's temperature and maintaining a habitable climate. These gases include carbon dioxide (CO_2), methane (CH_4), and nitrous oxide (N_2O). However, human activities, such as the burning of fossil fuels like coal, oil, and gas, have significantly increased their concentration in our atmosphere. This unprecedented rise in GHG levels is driving the phenomenon we know as global warming or simply, climate change, which is rapidly increasing the planet's temperature and placing our future at risk.

Major Greenhouse Gases (GHGS) Associated with Human Activities

Greenhouse gas	How it's produced	Average lifetime in the atmosphere	100-year global warming potential (GWP)
Carbon dioxide	Emitted primarily through the burning of fossil fuels (oil, natural gas, and coal), solid waste, and trees and wood products. Changes in land use also play a role. Deforestation and soil degradation add carbon dioxide to the atmosphere, while forest regrowth takes it out of the atmosphere.	Reference gas (see below*)	1*
Methane	Emitted during the production and transport of oil and natural gas as well as coal. Methane emissions also result from livestock and agricultural practices and from the anaerobic decay of organic waste in municipal solid waste landfills.	11.8 years	27.0–29.8**
Nitrous oxide	Emitted during agricultural and industrial activities, as well as during combustion of fossil fuels and solid waste.	109 years	273
Fluorinated gases	A group of gases that contain fluorine, including hydrofluorocarbons, perfluorocarbons, and sulfur hexafluoride, among other chemicals. These gases are emitted from a variety of industrial processes and commercial and household uses and do not occur naturally. Sometimes used as substitutes for ozone-depleting substances such as chlorofluorocarbons.	A few weeks to thousands of years	Varies (the highest is sulfur hexafluoride at 25,200)

This table shows 100-year global warming potentials, which describe the effects that occur over a period of 100 years after a particular mass of a gas is emitted. Global warming potentials and lifetimes come from Tables 7.15 and 7.SM.7 of the Intergovernmental Panel on Climate Change's Sixth Assessment Report, Working Group I contribution.

* Carbon dioxide's lifetime cannot be represented with a single value because the gas is not destroyed over time, but instead moves among different parts of the ocean–atmosphere–land system. Some of the excess carbon dioxide is absorbed quickly (for example,

14

by the ocean surface), but some will remain in the atmosphere for thousands of years, due in part to the very slow process by which carbon is transferred to ocean sediments.

** Methane's global warming potential is shown as a range that includes methane from both fossil and non-fossil sources.

Global Warming Potentials

Understanding Global Warming Potentials (GWPs) is crucial for grasping how different greenhouse gases (GHGs) contribute to Earth's warming. GHGs act like a blanket by absorbing heat from the Earth's surface and preventing it from escaping into space. This slows the loss of heat and warms the planet. While this process is natural, human activities are increasing GHG levels, intensifying global warming.

The impact of each GHG on warming varies based on two main factors: its ability to absorb energy, known as radiative efficiency, and its atmospheric lifetime.

To compare the warming impacts of different gases, scientists developed a metric known as Global Warming Potential (GWP). This metric measures how much energy the emissions of one ton of a gas will absorb over a specific period, usually 100 years, relative to one ton of carbon dioxide (CO_2). The higher the GWP, the greater the warming effect compared to CO_2 over that period. By using GWPs, we can standardize measurements, compile GHG inventories, and compare emissions reduction opportunities across different gases and sectors.

Let's look at some key greenhouse gases and their GWPs:

- **Carbon dioxide** (CO_2) has a GWP of 1, serving as the reference gas. Its atmospheric lifetime spans thousands of years, making its impact significant over millennia.

- **Methane** (CH_4) has a GWP of 27-30 over 100 years and a lifetime of about a decade. Despite its shorter lifetime, methane's higher energy absorption and indirect effects, such as its role as an ozone precursor, contribute to its higher GWP.

- **Nitrous oxide** (N_2O) has a GWP of 273 over 100 years and an atmospheric lifetime exceeding 100 years. Its long-lasting presence and high energy absorption make it a particularly potent GHG.

- **Other high-GWP gases** include chlorofluorocarbons, hydrofluorocarbons, hydrochlorofluorocarbons, perfluorocarbons, and sulfur hexafluoride. These gases have GWPs in the thousands or tens of thousands, meaning they trap substantially more heat than CO_2 for a given mass.

Understanding GWPs enables policymakers and analysts to prioritize and strategize emissions reduction efforts effectively across various sectors and gases. By doing so, they can make informed decisions that contribute to mitigating the effects of climate change.

Now, let's visualize this rise in greenhouse gas concentrations and examine how significantly they have increased, as well as the continuing trend, in the following diagram:

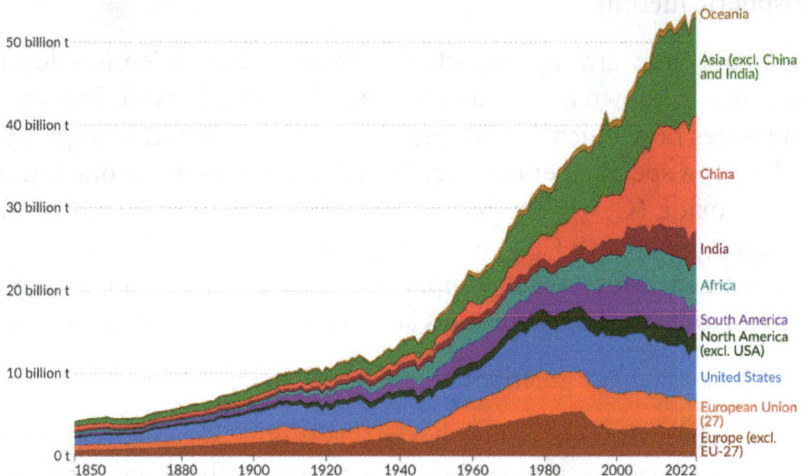

Annual greenhouse gas emissions by world region, 1850 to 2022

Greenhouse gas emissions[1] include carbon dioxide, methane and nitrous oxide from all sources, including land-use change. They are measured in tonnes of carbon dioxide-equivalents[2] over a 100-year timescale.

As this graph shows, the upward trend in greenhouse gas concentrations is closely correlated with the industrial revolution and the increased use of fossil fuels. This rapid buildup of greenhouse gases is trapping more heat in the atmosphere, leading to a rise in global temperatures.

Let's take a closer look at some key numbers to understand the magnitude of this issue. Since the preindustrial era, the concentration of CO_2 in the atmosphere has increased by about 47%. This alarming rise is primarily attributed to the burning of fossil fuels for energy production, transportation, and industrial processes.

Methane, another potent greenhouse gas, has accounted for around 16% of global warming since preindustrial times. It is released from sources such as livestock farming, fossil fuel extraction, and natural wetlands. Over the past two centuries, methane concentrations have more than doubled.

Nitrous oxide, mainly emitted from agricultural and industrial activities, has also significantly contributed to global warming. Its concentration has increased by approximately 23% since preindustrial times and is now significantly higher than natural background levels.

These consequences of global warming are not just theoretical—they are already unfolding before our eyes, and the effects are alarming. Rising global temperatures are causing shifts in weather patterns, leading to more frequent and intense heatwaves, droughts, wildfires, floods, tornadoes, and storms. These extreme weather events pose severe risks to human lives, infrastructure, and ecosystems.

One of the primary contributors to global warming is the power sector, which heavily relies on fossil fuels for electricity generation. Burning these fuels releases vast quantities of CO_2 and other greenhouse gases into the atmosphere, disrupting the delicate balance of our planet's climate system.

These numbers and trends underscore the urgent need to address greenhouse gas emissions and combat global warming. Transitioning to low-carbon energy sources, including nuclear power, is essential to reducing these emissions and mitigating the effects of climate change.

This is a critically important problem because the excessive buildup of greenhouse gases in the atmosphere is leading to the greenhouse effect, trapping heat from the sun and causing the Earth's temperature to rise. This phenomenon has serious consequences for our environment and our future.

Sea Level Rise or Glug! Glug!

Sea level rise is no longer a distant threat—it is happening now, causing coastal erosion and flooding in many areas. Low-lying coastal regions and small island nations are particularly at risk, facing the dual threat of submersion and displacement. Rising sea levels not only increase the risk of storm surges but also intensify the damage caused by hurricanes and other extreme weather events. Moreover, saltwater intrusion into freshwater sources is becoming a critical issue, threatening drinking water supplies and undermining agricultural productivity.

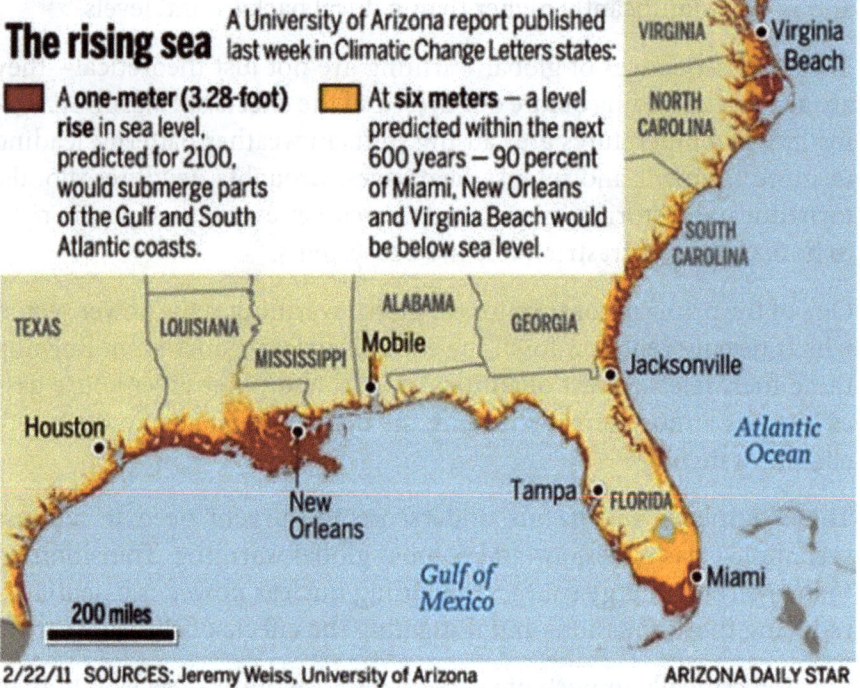

The rising sea A University of Arizona report published last week in Climatic Change Letters states:

A **one-meter (3.28-foot) rise** in sea level, predicted for 2100, would submerge parts of the Gulf and South Atlantic coasts.

At **six meters** — a level predicted within the next 600 years — 90 percent of Miami, New Orleans and Virginia Beach would be below sea level.

VIRGINIA · Virginia Beach

NORTH CAROLINA

SOUTH CAROLINA

TEXAS   LOUISIANA

ALABAMA   GEORGIA

MISSISSIPPI   Mobile

Jacksonville

Houston

Atlantic Ocean

Tampa · FLORIDA

New Orleans

Gulf of Mexico

· Miami

200 miles

2/22/11 SOURCES: Jeremy Weiss, University of Arizona   ARIZONA DAILY STAR

South Florida, renowned for its beautiful beaches and vibrant communities, is now facing a pressing threat from rising sea levels. Over the past several decades, this region has experienced a noticeable increase, exacerbated by global warming.

With its low-lying topography and extensive coastline, South Florida is particularly vulnerable to these changes. Sea levels have risen about a foot over the past 80 years, with 8 inches of that increase occurring in the last 30 years. This acceleration is expected to continue, with projections indicating that the next foot of sea level rise could occur within just 30 years, and subsequent feet in even shorter intervals.

The impact of rising sea levels in South Florida is multifaceted and severe. Coastal erosion is accelerating, leading to the loss of valuable land and infrastructure. Saltwater intrusion into freshwater aquifers threatens the region's water supply, posing significant risks to agriculture and drinking water. Additionally, the frequency and severity of flooding events have increased, affecting communities, businesses, and ecosystems leading to skyrocketing costs for home, property, and auto insurance.

South Florida is the 'Canary in the Coal Mine' of Climate Change... sun-sentinel.com

Miami, often cited as ground zero for sea level rise, has already invested in substantial flood mitigation efforts, such as elevating roads and installing water pumps. However, these measures are seen as temporary fixes in the face of a relentless and accelerating threat.

Looking into the future, the effects of sea level rise will become even more pronounced. According to the Intergovernmental Panel on Climate Change (IPCC), global mean sea levels could rise by up to one meter by the end of this century, depending on the rate of greenhouse gas emissions. However, it's important to note that predicting the exact extent of sea level rise is challenging due to the complexity of the

Earth's climate system. The IPCC has a Sea Level Projection tool that provides consensus projections on future sea levels across the globe under a range of possible future scenarios: https://sealevel.nasa.gov/ipcc-ar6-sea-level-projection-tool.

The consequences of future sea level rise will include the displacement of millions of people who live in coastal areas and on islands. This will lead to population migration, strained infrastructure, and increased pressure on resources in inland regions. For island nations, the effects are particularly severe, as entire communities may be forced to relocate due to rising waters, eroding shorelines, and loss of land. Many island populations, already vulnerable due to limited resources, face existential threats, including the potential loss of cultural heritage and sovereignty as they seek refuge in other countries. Economically, coastal cities and industries such as tourism, fishing, and shipping will be heavily impacted. The loss of coastal habitats like wetlands and coral reefs will also have ecological consequences, affecting biodiversity and ecosystem functioning. It also has implications for agriculture, as changing weather patterns can impact crop yields and food security.

The Niño and Niña Twins

El Niño and La Niña are two natural climate phenomena with significant global impacts. El Niño occurs when there is a warming of ocean surface temperatures in the Equatorial Pacific, while La Niña results from a cooling of these temperatures. These variations can have far-reaching effects on weather patterns and climate systems worldwide.

During El Niño events, the warming of the ocean surface leads to changes in atmospheric circulation patterns. This shift can cause increased rainfall in certain regions, such as the southwestern United States and parts of South America, while other areas suffer drought conditions. El Niño can also influence the intensity and frequency of tropical storms and hurricanes, often leading to more active Atlantic hurricane seasons.

On the other hand, La Niña events have effects opposite to those of El Niño. The cooling of ocean surface temperatures cools the atmosphere, reducing water evaporation and resulting in drier conditions in some areas. This can lead to droughts and an increased risk of wildfires. La Niña can also impact the intensity and tracks of tropical storms, often leading to fewer Atlantic hurricanes but more in the Pacific.

Both El Niño and La Niña can have significant impacts on agriculture, water resources, and ecosystems. Changes in rainfall patterns and temperature can affect crop yields, causing food shortages and economic losses. They can also disrupt marine ecosystems, impacting fisheries and coral reef health. Moreover, the effects of El Niño and La Niña can extend beyond their immediate regions, influencing global weather patterns and climate variability.

It is important to note that while El Niño and La Niña events are natural climate phenomena, evidence suggests that their frequency and intensity may be influenced by climate change. As the Earth's climate continues to warm, it is expected that the impacts of these events will become more pronounced and potentially more unpredictable. Understanding and monitoring these phenomena is crucial for developing effective climate change adaptation and mitigation strategies.

The Human and Economic Costs of Climate Change

Climate change is a global crisis that poses severe threats to both human well-being and economic stability. As the planet continues to warm due to greenhouse gas emissions, we are witnessing a wide range of adverse impacts that have significant consequences for society as a whole. Here we explore the current and future human and economic costs of climate change, shedding light on the urgency of addressing these pressing issues.

These costs encompass not only the financial burden but also the heightened risks to our health. Climate change not only worsens existing health problems but also introduces new ones. According to the World Health Organization (WHO), the estimated health costs of climate change are projected to reach $2 to 4 billion per year by 2030. Rising temperatures contribute to an increase in heat-related illnesses like heatstroke and dehydration. Additionally, changing weather patterns and expanding ranges for disease-carrying insects facilitate the spread of infectious diseases such as malaria and dengue fever.

A study published in Nature Communications highlights a critical future scenario where heat exposure for older adults, who are more vulnerable to high temperatures, will significantly increase due to global warming. By 2050, the number of older individuals exposed to dangerous heat levels (37.5°C, or 99.5°F, and above) will at least double worldwide, adding 250 million people or more aged 69 or above to those facing hazardous heat conditions. Health systems will be severely impacted, and since the most affected regions are typically poorer and already hot, this will add to global inequality.

The treatment of these heat-related illnesses, vector-borne diseases, and mental health issues resulting from climate-related disasters will strain our healthcare systems, leading to increased expenditure. In a study published in The Lancet, a renowned medical journal, climate change is called "code red for a healthy future", and they estimate an even higher economic cost of health damage caused by climate change than those estimated by the World Health Organization.

Changing rainfall patterns and extreme weather events disrupt agricultural systems, leading to crop failures and food shortages. The Food and Agriculture Organization (FAO) estimates that the economic costs of climate change-related impacts on agriculture could reach $18-32 billion per year by 2050. Rising sea levels and increased salinity of freshwater sources threaten access to clean drinking water for many millions of people living in coastal areas around the world.

As climate change intensifies, vulnerable communities will be forced to relocate due to sea-level rise, extreme weather events, and environmental degradation. This displacement will put immense strain on social structures and resources, leading to increased conflicts and humanitarian crises. Additionally, the continued melting of glaciers and ice sheets will further contribute to sea-level rise, posing a significant threat to low-lying coastal areas and small island nations. The World Bank estimates that by 2050, there could be 140 million climate migrants within regions of Sub-Saharan Africa, South Asia, and Latin America. Moreover, the economic costs of sea-level rise alone are projected to reach $1 trillion per year by 2050, according to the World Bank.

Climate-related disasters such as hurricanes, floods, and wildfires cause significant damage to infrastructure, including roads, bridges, buildings, and power grids. The United Nations Office for Disaster Risk Reduction (UNDRR) estimates that the annual economic losses from climate-related disasters could surpass $500 billion by 2030. The cost of repairing and rebuilding these structures is a substantial burden on national economies.

Changes in temperature and precipitation patterns disrupt agricultural systems, leading to reduced crop yields and increased production costs. The Intergovernmental Panel on Climate Change (IPCC) estimates that the economic costs of climate change on global agriculture could range from $17 to 35 billion per year by 2050. This not only impacts food security but also affects the livelihoods of millions of farmers and agricultural workers worldwide.

Climate change is expected to intensify the frequency and severity of extreme weather events such as hurricanes, droughts, and wildfires. These events will result in significant economic losses, damage to infrastructure, and loss of life and livelihoods. The Global Commission on Adaptation estimates that the economic costs of climate change-related disasters could reach $9 trillion by 2030.

Climate change disproportionately affects vulnerable communities and worsens existing socioeconomic disparities. The poorest countries and marginalized populations within them bear the brunt of climate-related impacts, widening the gap between the rich and the poor. According to Oxfam, the economic costs of climate change could push an additional 100 million people into poverty by 2030.

The human and economic costs of climate change are substantial and demand urgent action on a global scale. Mitigation efforts, such as reducing greenhouse gas emissions and transitioning to renewable energy sources, are crucial to limit the severity of future impacts. Adaptation measures, including investing in resilient infrastructure, developing early warning systems, and supporting vulnerable communities, are essential for minimizing the human suffering and economic losses associated with climate change. By recognizing the interconnectedness of climate change with human well-being and economic stability, we can work towards a sustainable and resilient future for all.

Additional Effects on our Everyday Life

On CBS Face the Nation on July 9, 2023, NASA Chief Scientist and Climate Advisor, Kate Calvin, PhD, discussed how climate change is already affecting everyday life. Some effects she mentioned include:

- Higher incidence of severe turbulence for airline flights
- Cargo shipments impacted by river levels
- Increased influence of Lyme disease, malaria, and other mosquito-borne diseases
- Seasonal allergies worsening
- More fire weather and lengthened fire seasons
- More serious air quality problems

These effects are in addition to significantly increasing economic costs for all types of insurance.

Conclusion

The current problem we face is the impact of greenhouse gas emissions on our planet. Global warming is causing severe climate disruptions and posing risks to our environment and costs to society.

By embracing nuclear power, a low-carbon energy source, we can significantly reduce our dependence on fossil fuels and decrease the amount of CO_2 and other greenhouse gas emissions, paving the way for a sustainable future. This transition can help mitigate the effects of global warming and combat climate change. Moreover, nuclear power provides a reliable and constant source of electricity, ensuring a stable energy supply for our growing needs.

Of course, it's important to address concerns regarding nuclear power, such as safety, waste management, and proliferation, and I'll do that in subsequent chapters. However, with advancements in technology and strict regulatory frameworks, these challenges can be overcome. Keep in mind that nuclear power must be added to an expansion of all our other renewable energy sources to create a well-rounded and sustain-able energy mix.
In chapter 3 we'll examine the status of these solutions, exploring their potential to replace fossil fuels and reduce emissions.

References

1. *Climate change indicators: Greenhouse gases | US EPA.* (2024, June 27). US EPA. https://www.epa.gov/climate-indicators/greenhouse-gases
2. *The Paris Agreement - Publication | UNFCCC.* (n.d.). https://unfccc.int/documents/184656
3. *Global Climate Change: Evidence.* (2008, June 15). Retrieved January 14, 2015, from http://climate.nasa.gov/evidence/
4. *Global warming of 1.5 ºC—.* (n.d.). Global Warming of 1.5 ºC. https://www.ipcc.ch/sr15/
5. Ritchie, H., Rosado, P., & Roser, M. (2023, December 28). *CO_2 and Greenhouse Gas Emissions.* Our World in Data. https://ourworldindata.org/co2-and-greenhouse-gas-emissions#all-charts
6. *IPCC AR6 Sea Level Projection Tool.* (n.d.). NASA Sea Level Change Portal. https://sealevel.nasa.gov/ipcc-ar6-sea-level-projection-tool
7. Romanello, M., McGushin, A., Di Napoli, C., Drummond, P., Hughes, N., Jamart, L., Kennard, H., Lampard, P., Rodriguez, B. S., Arnell, N., Ayeb-Karlsson, S., Belesova, K., Cai, W., Campbell-Lendrum, D., Capstick, S., Chambers, J., Chu, L., Ciampi, L., Dalin, C., . . . Hamilton, I. (2021). *The 2021 report of the Lancet Countdown on health and climate change: code red for a healthy future. Lancet, 398*(10311), 1619–1662. https://doi.org/10.1016/s0140-6736(21)01787-6
8. *2023: A historic year of U.S. billion-dollar weather and climate disasters.* (2024, January 8). NOAA Climate.gov. https://www.climate.gov/news-features/blogs/beyond-data/2023-historic-year-us-billion-dollar-weather-and-climate-disasters
9. *World Energy Outlook 2020 – Analysis - IEA.* (2020, October 1). IEA. https://www.iea.org/reports/world-energy-outlook-2020

Chapter 3

The Current Status of Mitigation Approaches: Evaluating Renewables and Fossil Fuels

Let's start this chapter by looking at where we in the U.S. currently get the power we need to sustain our standard of living:

U.S. primary energy consumption by energy source, 2021

total = 97.33 quadrillion British thermal units (Btu)

total = 12.16 quadrillion Btu

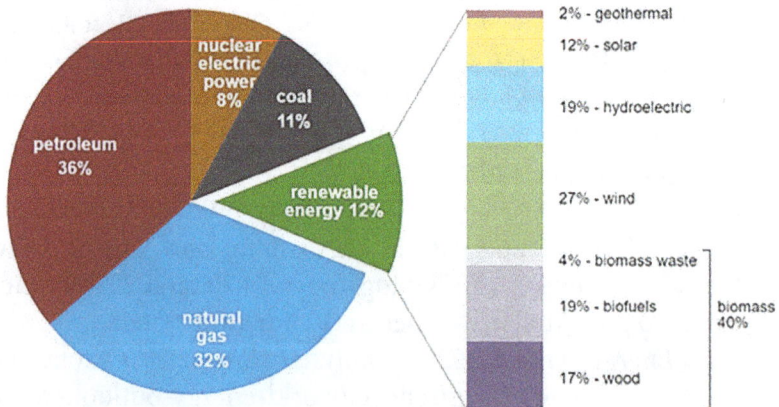

petroleum 36%

nuclear electric power 8%

coal 11%

renewable energy 12%

natural gas 32%

2% - geothermal

12% - solar

19% - hydroelectric

27% - wind

4% - biomass waste

19% - biofuels

17% - wood

biomass 40%

Data source: U.S. Energy Information Administration, *Monthly Energy Review*, Table 1.3 and 10.1, April 2022, preliminary data
Note: Sum of components may not equal 100% because of independent rounding.

eia

Even though the renewable energy portion has been growing in the U.S., as of the latest data, it still accounts for only about 12% of our energy needs. Adding nuclear electric power at 8%, we are still heavily dependent on fossil fuels, which supply 80% of our power and continue to emit vast amounts of greenhouse gases.

In 2023, global energy consumption was still overwhelmingly dominated by fossil fuels, which accounted for approximately 81.5% of the total primary energy demand. Although this marked a slight decrease from previous years, fossil fuels remained the majority source of energy worldwide. Renewable energy sources, including wind, solar, and hydropower, contributed about 14.6% of global energy consumption. This share has been growing rapidly, with renewables expanding at a pace significantly faster than the overall growth in primary energy consumption. Nuclear energy maintained a relatively stable contribution, accounting for around 4% of the total global energy mix.

Let's now delve into the current status of mitigation approaches in combating global warming, focusing specifically on fossil fuels and renewable energy sources. By understanding the strengths and limitations of each, we can gain valuable insights into how nuclear power, with its unique attributes, can play a significant role in mitigating climate change.

Fossil Fuels and Carbon Capture

Despite the growing adoption of renewable energy sources, fossil fuels continue to dominate the global energy landscape. These longstanding energy sources have powered human civilization for centuries, but their use is a major driver of climate change. The combustion of fossil fuels releases vast amounts of carbon dioxide (CO_2), methane, and other pollutants into the atmosphere, exacerbating global warming and contributing to severe air pollution.

Carbon Capture, Utilization, and Storage (CCUS)

One potential solution to fossil fuel emissions is carbon capture, utilization, and storage (CCUS). This technology captures CO_2 as it is emitted from large sources like power plants and industrial facilities that burn fossil fuels or biomass. After capture, the CO_2 is compressed and transported via pipelines, ships, or trucks for either utilization in products or long-term storage in deep geological formations, such as depleted oil fields or saline aquifers. CCUS aims to keep CO_2 out of the atmosphere and reduce its impact on climate change.

While CCUS holds promise, large-scale deployment has been slow due to high costs and technical challenges. However, ongoing research and innovation may drive down these costs, making the technology a more viable solution in the future.

Negative Emission Technologies (NETs): The Next Frontier

Beyond capturing emissions from current sources, there is growing recognition that reducing atmospheric CO_2 levels also requires removing existing carbon through negative emission technologies. These methods actively draw CO_2 from the atmosphere, providing a path to not only reduce emissions but reverse some of the damage already done.

Some key NETs include:

- **Direct Air Capture (DAC):** DAC extracts CO_2 directly from the ambient air using solid sorbents or liquid solvents, which are then processed to capture the gas. The CO_2 can be stored underground or repurposed in industrial products. While promising, DAC is currently expensive, with costs ranging from \$250 to \$600 per metric ton of CO_2 and requires significant energy to operate. Nevertheless, companies and governments are pushing for innovations to make DAC more cost-effective.

- **Reforestation and Afforestation:** Planting trees and restoring forests help absorb CO_2 from the atmosphere through photosynthesis. Unlike reforestation, which involves replanting trees in areas that were previously forested, afforestation establishes forests in places that have not historically had tree cover. These methods offer a natural carbon removal strategy.

- **Bioenergy with Carbon Capture and Storage (BECCS)**: BECCS involves burning biomass to generate energy while capturing and storing the CO_2 emissions. This approach combines renewable energy production with carbon sequestration.

- **Soil Carbon Sequestration**: Agricultural practices that promote soil health, such as no-till farming and composting, can help store carbon in the soil, reducing CO_2 levels in the atmosphere.

These technologies are vital for achieving net-zero emissions, where the amount of CO_2 removed from the atmosphere equals or exceeds the amount being emitted.

Carbon Removal Credits: Incentivizing Action

To support the development and scaling of negative emission technologies, carbon removal credits have emerged as a financial mechanism. Each credit represents the removal of one metric ton of CO_2 from the atmosphere, verified through independent audits to ensure the removal is measurable, additional, and permanent. These credits are traded on carbon markets, offering companies and individuals a way to offset their emissions while supporting projects that actively reduce atmospheric CO_2.

Projects generating these credits range from reforestation to advanced methods like direct air capture and biochar production. As more corporations commit to net-zero goals, the demand for carbon removal credits is growing, creating incentives for innovation and investment in these critical technologies.

Biochar: A Multifaceted Tool for Carbon Sequestration

Biochar is one of the more accessible and affordable negative emission technologies. Produced by heating organic materials (such as agricultural waste or wood chips) in a low-oxygen environment, biochar stores carbon in a stable form. When applied to soil, it can sequester carbon for decades or even centuries, removing CO_2 from the atmosphere.

Beyond carbon capture, biochar offers agricultural benefits. It improves soil fertility, enhances water retention, and reduces the need for synthetic fertilizers, promoting sustainable farming practices. While biochar's effectiveness depends on factors such as the type of biomass used and local soil conditions, ongoing research is exploring ways to maximize its carbon capture potential and ecological benefits.

One of biochar's main advantages is cost-effectiveness. As of 2023, biochar carbon removal credits cost between \$131 and \$220 per metric ton, significantly lower than other technologies like direct air capture. Major corporations, including Microsoft and Shell, are investing in biochar projects, helping to drive market growth and scale its implementation.

A Path Forward

Fossil fuels have been foundational to global energy systems, but their continued use poses immense risks to the environment and human health. In addition to contributing to climate change, the extraction, transport, and combustion of fossil fuels have significant ecological and economic costs. As fossil fuels are finite, their eventual scarcity will make them economically unsustainable.

Negative emission technologies, along with carbon capture and storage, provide a critical pathway for mitigating these risks. While still in their early stages, technologies like direct air capture, biochar, and soil carbon sequestration offer promising solutions to reduce and even reverse the harmful effects of greenhouse gas emissions. By coupling these innovations with financial incentives like carbon removal credits, we can chart a course toward a more sustainable and climate-resilient future.

Renewable Energy Sources

Renewable energy sources, including solar, wind, hydroelectric, and geothermal power, have gained considerable attention in recent years. They are touted as clean and sustainable alternatives to fossil fuels. These sources harness naturally occurring energy flows and emit minimal greenhouse gases during operation.

Share of primary energy consumption from renewable sources, 2023
Measured as a percentage of primary energy¹ using the substitution method². Renewables include hydropower, solar, wind, geothermal, bioenergy, wave, and tidal, but not traditional biofuels, which can be a key energy source, especially in lower-income settings.

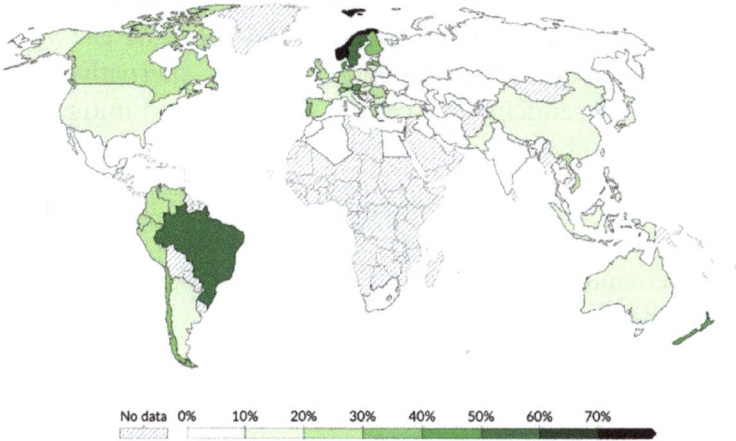

Solar Power

Solar power has made remarkable strides thanks to incredible progress in photovoltaic (PV) technology. This innovative technology harnesses the power of the sun to generate electricity, paving the way for a cleaner and more sustainable energy source. As technology advances and costs decrease, solar power is becoming increasingly appealing to individuals, businesses, and communities. However, despite significant reductions in the cost of solar panels, solar energy remains more expensive than traditional fossil fuels, limiting accessibility, particularly in developing countries.

Solar power faces significant challenges due to its intermittent nature. Since it relies on weather conditions, it isn't available 24/7 like traditional fossil fuels or nuclear power. During cloudy days or at night, energy production from solar panels is reduced or halted. To counter this, energy storage solutions have been developed to ensure a consistent power supply even when the sun isn't shining. These solutions store excess energy generated during sunny periods, which can then be tapped during less sunny times, ensuring a continuous power supply for homes, businesses, and the electricity grid. Ongoing research is focused on improving the efficiency and affordability of storage batteries.

The beauty of solar power lies in its ability to tap into an abundant and renewable energy source: the sun. With the right infrastructure and continued technological advancements, solar power has the potential to revolutionize how we generate and consume electricity. It offers a sustainable and environmentally friendly alternative to fossil fuels, significantly reducing our carbon emissions and mitigating climate change.

However, a significant challenge remains: the expanding footprint of solar arrays. As cities increasingly turn to solar arrays to meet their electricity demands, the amount of land needed for these large-scale installations becomes a pressing concern.

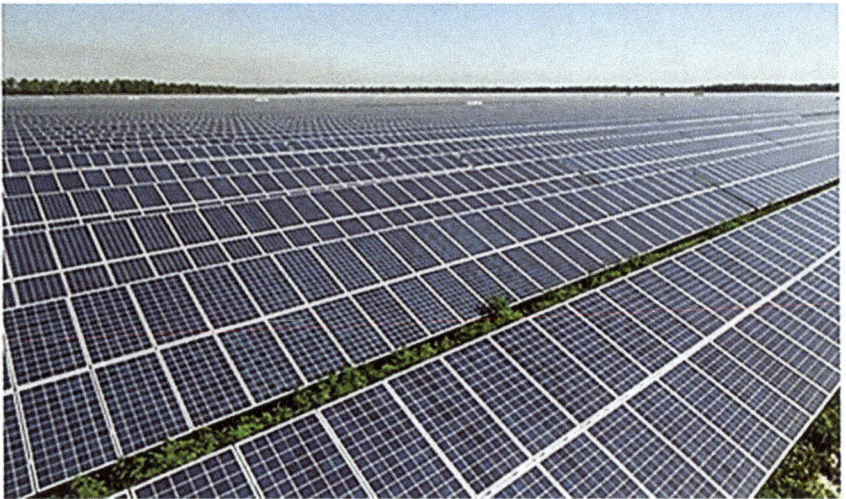

A Large Solar Array   *Florida Insider*

To supply electricity to cities, solar arrays require a considerable amount of land. The exact land footprint depends on various factors, including the capacity of the solar installation and the efficiency of the panels. On average, a utility-scale solar photovoltaic (PV) power plant requires approximately 3-5 acres of land per megawatt (MW) of capacity. This means that to generate 100 MW of electricity, around 300-500 acres of land would be needed. Considering that a typical coal-fired power plant in the United States has a capacity ranging from 500 to 1,500 megawatts (MW), replacing that plant with a solar array might take from 5 to 15 times as much land area. That's a lot of space!

As solar technology continues to advance, the land requirements for solar arrays are expected to decrease. The efficiency of solar panels is steadily improving, allowing more electricity to be generated from a smaller surface area. Additionally, innovations like floating solar arrays and solar panels integrated onto buildings and infrastructure are reducing the need for large tracts of land. These developments maximize the utilization of available space, such as rooftops, parking lots, and water bodies.

For densely populated cities, finding vast expanses of land for solar arrays may simply not be possible. However, urban solar integration presents an exciting opportunity to utilize existing infrastructure for solar power generation. Rooftop solar installations can capitalize on unused space on buildings and contribute to the overall energy supply of the city. Moreover, solar canopies and solar facades can be integrated into urban landscapes, further optimizing land use.

While solar arrays require significant land to generate electricity for cities, ongoing advancements in technology and innovative approaches to urban solar integration are expected to reduce their land footprint. As we continue to invest in solar energy and explore new possibilities, it's crucial to balance meeting our energy needs and preserving our natural resources. By embracing solar power and optimizing land use, we can help pave the way toward a more sustainable and greener future for our cities.

Personal Note: My husband and I spent years sailing around the world on small yachts. For most of our voyage, we had two solar panels mounted on the stern pulpit of our yacht, and they produced most of the power we needed at anchor and on long passages. Solar panels are significantly more efficient and cheaper now.

Wind Energy

Wind farms have seen remarkable growth, emerging as the leading source of renewable energy globally. Harnessing the kinetic energy of the wind through turbines to generate electricity has proven to be a reliable and sustainable solution.

However, wind energy is not without its challenges. One of the primary limitations is the variability of wind resources. Wind speeds can fluctuate significantly depending on the time of day, season, and geographical location. This variability poses difficulties in integrating wind energy into the power grid and ensuring a consistent supply of electricity.

To address this issue, advanced forecasting techniques are employed to predict wind patterns, enabling better grid management and energy balancing. Additionally, as with solar energy, the development of energy storage technologies—such as batteries and pumped hydro storage—allows for the storage of excess energy during high wind periods for use during low wind periods, further enhancing grid stability.

Another significant concern associated with wind energy is its impact on wildlife, particularly birds and bats. The rotating blades of wind turbines can pose a risk of collision, leading to injury or mortality among avian and bat populations. However, extensive research and technological advancements have focused on mitigating these risks. Careful turbine placement and the implementation of curtailment strategies during migration seasons have proven effective in reducing bird and bat collisions. Ongoing studies aim to enhance turbine designs, such as optimizing blade shapes and colors, to further minimize their impact on wildlife.

The size of wind turbines also presents a challenge. As turbines increase in height, they become more efficient at capturing wind energy. However, larger turbines also introduce challenges in terms of cost, transportation, and installation. To overcome these limitations, innovations are being pursued. For example, the development of lighter and stronger materials, such as carbon fiber, reduces the overall weight of the turbines, making transportation and installation more feasible. Offshore wind farms, which benefit from stronger and more consistent winds, offer opportunities for larger turbine deployment, thereby increasing energy generation potential.

To address concerns related to wind energy, comprehensive environmental impact assessments and careful planning are essential. Government regulations and policies play a crucial role in ensuring responsible wind farm development. Environmental organizations and stakeholders are actively involved in shaping these policies to strike a balance between renewable energy goals and environmental conservation.

As wind energy continues to evolve, ongoing research and development efforts are focused on addressing these limitations and maximizing its potential as a sustainable energy source. By finding innovative solutions to overcome variability, mitigating wildlife impacts, and optimizing turbine design and deployment, wind energy can further contribute to the global transition toward a clean and renewable energy future.

Hydroelectric Energy

Hydroelectric power, or hydropower, has been a cornerstone of renewable energy for many years, derived from harnessing the power of flowing or falling water. It generates electricity by converting the kinetic energy of moving water into power, typically through the construction of dams or diversion structures that control the flow of rivers or other bodies of water. The water flows through turbines, which spin and drive generators to produce electricity. The amount of energy generated depends on the volume of water flow and the change in elevation, or head, from one point to another.

Photo by T L on Unsplash

Hydropower offers the advantage of providing consistent power generation. However, as with solar and wind energy, we must acknowledge the limitations associated with this form of energy. A primary limitation of hydroelectric energy is its reliance on water availability. Hydroelectric power plants require a steady supply of water to generate electricity efficiently, making them vulnerable to variations in precipitation levels, droughts, or other water scarcity issues. In regions experiencing water shortages or drought conditions, hydroelectric power generation may be limited. It is crucial to manage water resources sustainably to mitigate the impact of this limitation.

Another significant concern associated with hydroelectric energy is its impact on local ecosystems. The construction of large-scale hydroelectric dams can alter river ecosystems, affecting water flow, sediment transport, and habitat integrity. These changes can have adverse effects on fish populations, migratory patterns, and other wildlife dependent on the river ecosystem. To address these concerns, environmental impact assessments and the implementation of mitigation measures are essential. For instance, fish ladders or fish bypass systems can be integrated into dam designs to facilitate fish migration and minimize disruptions to aquatic biodiversity.

Furthermore, the construction of hydroelectric dams can cause significant population displacement. Communities living in areas designated for dam construction often face relocation, disrupting their livelihoods and cultural ties to the land. The social and economic impact on displaced populations can be profound, necessitating careful planning, fair compensation, and support for resettlement to mitigate these adverse effects. Engaging with affected communities and ensuring their participation in the planning process is crucial for addressing these challenges.

Additionally, the availability of suitable sites for hydroelectric dams poses a limitation to the expansion of hydroelectric energy. Not all rivers and streams are suitable for dam construction due to factors such as topography, water flow characteristics, and environmental considerations. This limitation means that the potential for hydroelectric energy generation is constrained by the availability of suitable sites.

Despite these limitations, hydroelectric energy continues to play a significant role in global electricity production. Ongoing research and development efforts are focused on addressing these challenges and maximizing the potential of hydroelectric energy while minimizing its environmental impact. Innovative approaches, such as small-scale or run-of-the-river hydroelectric projects, are being explored to reduce ecosystem disruption and provide localized energy solutions.

It is important to note that the development of hydroelectric projects requires careful consideration of environmental, social, and economic aspects. Community engagement, proper planning, and the implementation of sustainable practices are crucial for minimizing potential negative impacts and ensuring the long-term viability of hydroelectric energy as a renewable energy source.

Tidal and Wave Energy

Other methods of producing electricity from moving water are being studied, including tidal energy that harnesses the energy produced by the rise and fall of ocean tides. It often uses turbines placed in tidal streams or barrages constructed across tidal estuaries. These systems convert the kinetic energy of moving water during tidal changes into electricity. Tidal energy is highly predictable due to the regularity of tidal cycles, but it is still in the early stages of development and has limited commercial application.

Wave energy captures kinetic energy from the surface motion of ocean waves. Various devices, such as oscillating water columns, point absorbers, and overtopping devices, are used to convert this motion into electricity. Wave energy systems are typically placed on or just below the water surface and can be anchored to the ocean floor. This form of energy is also in the early stages of development and faces challenges related to high costs and environmental impact.

Comparison Table: Hydroelectric, Tidal and Wave Energy

Aspect	Hydroelectric Power	Tidal Energy	Wave Energy
Energy Source	Flowing water in rivers or reservoirs	Rise and fall of ocean tides	Surface motion of ocean waves
Technology	Dams, diversion structures, turbines, generators	Tidal turbines, barrages, tidal lagoons	Oscillating water columns, point absorbers, overtopping devices
Predictability	High, based on water flow and elevation change	Very high, based on predictable tidal cycles	Moderate, based on wave patterns and wind

Aspect	Hydroelectric Power	Tidal Energy	Wave Energy
Development Stage	Mature, widely used	Early stages, limited commercial use	Early stages, limited commercial use
Environmental Impact	Potential displacement of communities, impact on aquatic ecosystems	Potential impact on marine life and tidal ecosystems	Potential impact on marine habitats, high costs

Geothermal Energy

Geothermal power, which harnesses the Earth's heat to generate electricity and provide heating, is a promising renewable energy source with several advantages, including a continuous and reliable power supply with low carbon emissions. However, it's essential to also consider the limitations associated with geothermal energy.

One of the primary limitations of geothermal energy is its dependence on specific geological features. Geothermal resources are typically found in regions with active volcanic activity or areas with hot springs and geysers. These geological features allow for the extraction of heat from the Earth's crust, but not all regions possess the necessary conditions, limiting the accessibility of geothermal energy. Despite this limitation, advancements in technology, such as enhanced geothermal systems (EGS), are being explored to expand the reach of geothermal energy by tapping into deeper and hotter rock formations.

Another consideration is the cost of developing geothermal energy projects. While geothermal energy becomes relatively inexpensive once a power plant is operational, the initial investment required for exploration and drilling can be substantial. This upfront cost can pose a challenge, particularly for countries or regions with limited financial resources. However, because the long-term operational costs of geothermal power plants are relatively low, they can be economically viable over time.

Furthermore, the development of geothermal power plants can impact local ecosystems. The drilling and extraction processes can disrupt underground ecosystems, potentially affecting local flora and fauna. It is crucial to conduct thorough environmental impact assessments and implement mitigation measures to minimize these effects. Additionally, sustainable practices and responsible management of geothermal reservoirs are essential to ensure the long-term sustainability of geothermal energy projects.

Despite these limitations, ongoing research and development efforts are focused on expanding the accessibility of geothermal energy. Technological advancements, such as enhanced geothermal systems and binary cycle power plants, are being pursued to tap into lower-temperature geothermal resources and increase the efficiency of power generation. Additionally, international collaborations and knowledge sharing are contributing to the growth and development of geothermal energy worldwide.

While geothermal energy may not be universally accessible, it holds significant potential in regions with suitable geological conditions. It is an important component of the renewable energy mix and offers a reliable and sustainable option for electricity generation and heating.

Bioenergy

Derived from organic matter like wood, crops, and waste, bioenergy is another renewable energy source with vast potential. However, as with other renewable energy sources, it is important to acknowledge the limitations associated with its utilization.

One of the primary limitations of bioenergy is its potential impact on food production. As bioenergy crops require land and resources, there is a concern that their cultivation may compete with food crops, leading to higher food prices and potential food scarcity, particularly in regions where food security is already a challenge. Careful planning and sustainable land management strategies are necessary to ensure a balanced approach that minimizes the impact on food production while promoting bioenergy development.

Another limitation of bioenergy is the potential environmental impact on local ecosystems. The cultivation of bioenergy crops, especially on a large scale, can result in the conversion of natural habitats and the loss of biodiversity. It is essential to consider the ecological implications and implement sustainable practices, such as agroforestry or mixed cropping systems, to minimize disruption to local ecosystems. The use of waste biomass for bioenergy production can help reduce environmental impacts by utilizing materials that would otherwise be discarded.

Furthermore, the efficiency of bioenergy conversion processes presents a limitation. Currently, bioenergy conversion technologies, such as biomass combustion or anaerobic digestion, have lower energy conversion efficiencies compared to other renewable sources like wind or solar energy. Ongoing research and development efforts in bioenergy focus on improving conversion technologies and exploring advanced biofuel production methods, such as bioethanol or biohydrogen, to enhance overall efficiency and energy output.

It is worth noting that bioenergy can provide several benefits, including reduced greenhouse gas emissions, waste management, and rural development opportunities. Integrated approaches, such as using residues or waste materials for bioenergy production, can maximize the sustainability and efficiency of bioenergy systems.

To overcome the limitations of bioenergy, comprehensive assessments and policies are necessary. These should consider factors such as land use planning, environmental safeguards, and social and economic implications. Furthermore, diversified renewable energy portfolios that combine bioenergy with other sources like wind, solar, and hydroelectric power can help mitigate limitations and ensure a reliable and sustainable energy supply.

As bioenergy technology evolves and sustainability practices are integrated, it has the potential to play a vital role in the transition to a low-carbon and more sustainable energy future.

How Green are Solar and Wind Power, Really?

Solar and wind power are often celebrated as the green heroes in our fight against climate change. Praised for their ability to drastically cut greenhouse gas emissions, they offer a promising alternative to fossil fuels. However, before we declare them the ultimate solution, it's crucial to look beyond the headlines and examine their full impact. These renewable energy sources are not without environmental and economic challenges. Understanding both their strengths and shortcomings is key to determining their true role in our energy future.

Solar Power: A Closer Look

Solar power is a crucial component of our renewable energy arsenal. It provides significant environmental benefits, chief among them its ability to reduce greenhouse gas emissions. Solar energy, unlike fossil fuels, doesn't pollute the air during electricity generation. This makes it a powerful tool in combating climate change and improving air quality.

However, the environmental benefits of solar power are tempered by drawbacks. Large-scale solar installations, especially in desert regions, can lead to land degradation and disrupt sensitive ecosystems. The production of solar panels is resource-intensive, requiring the mining of precious metals and significant energy consumption. Ironically, this process contributes to pollution and greenhouse gas emissions. Furthermore, solar panels have a limited lifespan of about 25-30 years, and with the lack of robust recycling processes, we face a growing challenge of e-waste as these panels reach the end of their life.

Another crucial aspect to consider is the need for large-capacity batteries to store the electricity generated by solar power. Solar energy is intermittent—it's only available when the sun is shining. To ensure a steady and reliable power supply, especially during the night or cloudy days, large-scale energy storage systems are necessary. These batteries, while essential for grid stability, come with environmental impacts. The production of batteries requires materials like lithium, cobalt, and nickel, involving energy-intensive mining processes that can lead to significant environmental degradation. Battery manufacturing and disposal also raise concerns about resource scarcity and toxic waste, adding complexity to the environmental footprint of solar power.

Economically, solar power has become more competitive. As of early 2023, the average cost for residential solar installations in the U.S. was about $3.28 per watt. While a significant portion of this cost comes from soft expenses like sales and permitting, the overall cost has decreased significantly, making solar power more accessible. Maintenance costs are low, and with solar panels lasting 25-30 years, they offer long-term savings and environmental benefits. The levelized cost of energy (LCOE), which factors in the total costs of building, operating, and maintaining a power plant over its expected lifetime, has also seen a steady decline, positioning solar power as a viable competitor to traditional energy sources.

Wind Power: An In-Depth Analysis

Wind power, like solar, is a key player in the shift towards renewable energy. It stands out as one of the cleanest energy sources available, generating electricity without emitting carbon dioxide or other harmful pollutants. Once wind turbines are installed, they require no fuel, relying instead on the natural, renewable power of the wind. This makes wind power a sustainable option with minimal ongoing environmental impact.

However, wind power is not without its challenges. The construction of wind turbines is resource-intensive, requiring significant amounts of steel, concrete, and rare earth metals. These materials have their own environmental footprints, tied to the extraction and processing required to produce them. Wind turbines can also impact the landscape, both visually and in terms of noise, which can be a concern for nearby communities. Additionally, while the risks are generally lower than those associated with fossil fuels, wind turbines can pose a threat to wildlife, particularly birds and bats.

Just as with solar power, the intermittent nature of wind energy necessitates the use of large-capacity batteries to store excess power generated during windy periods for use when the wind isn't blowing. The environmental implications of these batteries are similar to those for solar power, with concerns surrounding the mining of materials and the disposal of spent batteries. The lifecycle of batteries, from production to disposal, adds another dimension to the environmental impact of wind power that must be carefully managed.

From an economic perspective, wind power has seen substantial investment and growth. The cost of commercial wind turbines ranges from $2.6 to $4 million per turbine, with an average cost of about $1.3 million per megawatt of capacity. Operation and maintenance (O&M) costs, though significant, are manageable, averaging $42,000 to $48,000 per year per turbine. Over the lifetime of a turbine, these O&M costs can constitute 20-25% of the total levelized cost of electricity generated, making wind power a competitive option, particularly in regions with strong and consistent winds.

Weighing the Options and Navigating the Complexities

When comparing solar and wind power, both offer significant environmental benefits, particularly in reducing greenhouse gas emissions. Yet, they also come with unique challenges. Solar power's environmental impact is tied largely to its production process, land use, and the need for battery storage, while wind power's challenges are more related to its construction, visual impact, wildlife effects, and the need for batteries to ensure a stable energy supply.

Economically, both solar and wind have become increasingly viable options. Despite recent increases in commodity and financing costs, which have driven up prices by 10-15% since 2020, both remain among the most cost-effective sources for new electricity generation. The levelized cost of energy (LCOE) for both technologies continues to be competitive, making them attractive alternatives to fossil fuels.

Solar and wind power are critical allies in our quest for a sustainable energy future. They offer substantial environmental benefits, particularly in reducing greenhouse gas emissions and air pollutants. However, these benefits come with their own set of environmental and economic challenges. From the resource-intensive processes required to manufacture solar panels and wind turbines to the environmental impact of large-capacity batteries, these technologies demand careful consideration and management.

Despite these challenges, the long-term advantages of solar and wind power—low operational costs, minimal ongoing environmental impact, and their critical role in mitigating climate change—make them indispensable in our energy strategy. As technology continues to advance, including improvements in battery storage systems, these renewable energy sources are expected to become even more cost-effective and environmentally sustainable, solidifying their roles as cornerstones of our green energy future. (Please see Chapter 11 for more information on solar and wind LCOEs with attached energy storage).

Unlocking the Potential: 100% Clean Electricity by 2035

The National Renewable Energy Laboratory (NREL) study reveals an urgent and transformative opportunity: the United States can transition to 100% clean electricity by 2035, unlocking monumental climate and health benefits that far outweigh the costs. Funded by the U.S. Department of Energy's Office of Energy Efficiency and Renewable Energy, this study outlines a clear roadmap to decarbonize the U.S. power grid, underscoring that the actions we take today are critical to securing a sustainable future.

Key Findings That Demand Immediate Action:

1. **Multiple Viable Pathways**: NREL's scenarios—rooted in decades of cutting-edge research—show that a range of pathways can lead us to a net-zero power grid. These scenarios account for every variable, from the most optimistic projections of technological advancements to the most constrained conditions where certain technologies, like carbon capture, may not be feasible. In each case, the least-cost solution is modeled to ensure clean, reliable power 24/7, but the window for implementing these pathways is rapidly closing.

2. **Accelerated Deployment of Clean Energy**: By 2035, wind and solar energy could generate 60%–80% of the country's electricity, tripling today's total capacity. The scale required is nothing short of unprecedented. Without immediate and sustained action, we risk missing the opportunity to make this shift. If delays in siting and land use arise, nuclear power must step in, potentially more than doubling its capacity to keep us on track. The deployment of new technologies is not just a possibility— it is an absolute necessity, and it must happen now.

3. **Benefits Vastly Exceed Costs**: While the transformation of our energy system may cost between $330 billion to $740 billion, the climate and health benefits will dwarf these costs. Up to 130,000 lives could be saved from reduced air pollution, delivering between $390 billion to $400 billion in avoided mortality costs. On top of that, the U.S. could avoid more than $1.2 trillion in damage from floods, droughts, wildfires, and hurricanes—all of which are intensified by climate change. This is not just an investment in energy—it's an investment in the survival of our planet and the protection of our communities.

4. **Decisive Actions Needed Now**: The transition to 100% clean electricity demands swift and strategic action. We must drastically accelerate electrification, streamline the deployment of new infrastructure, and expand clean technology manufacturing and supply chains. The stakes are high: failure to act now could jeopardize these scenarios, making the transition significantly harder, more expensive, and less effective in the fight against climate change.

This study is a clarion call for immediate and decisive action. We have the technology, the pathways, and the knowledge to decarbonize our power grid by 2035. Every year of delay risks lives, economic stability, and the health of our planet. The findings make it clear: the time to act is now. Waiting any longer will not only increase the costs but also threaten the viability of reaching 100% clean electricity.

For nuclear energy, this presents a pivotal moment. If renewable technologies face deployment challenges, nuclear power must rise to the occasion—proving its critical role in providing clean, reliable energy. This is not just about reducing emissions; it's about securing the future of our planet, our economy, and the health of generations to come.

Conclusion: A Future of Possibilities

Renewable energy sources, particularly solar and wind, have made remarkable progress, becoming significantly cheaper and more accessible. The share of renewable energy in the global energy mix has more than tripled in recent years—a tremendous achievement. Yet, as we celebrate these successes, we must also recognize the challenges ahead.

Despite this growth, renewable energy still accounts for only a modest percentage of the world's energy production. To fully realize the potential of renewable energy, we must address the intermittent nature of these sources by developing more effective energy storage solutions. These systems will ensure a continuous and reliable power supply, even during periods of low renewable generation.

At the same time, we must leverage the potential of fossil fuel mitigation technologies, such as biochar and carbon capture, which are steadily improving. By capturing and storing carbon dioxide emissions from fossil fuel power plants, we can significantly reduce the carbon footprint associated with fossil fuel use.

The future of energy isn't about choosing one solution over another but finding the right combination that works. By integrating renewable energy sources, fossil fuel mitigation technologies, and clean nuclear power, we can bridge the gap between clean energy and reliable power supply. This balanced approach will set us on the path to a sustainable future, effectively mitigating the impacts of climate change.

The **100% Clean Electricity by 2035 Study** makes it clear: a decarbonized future is within our grasp, but only if we act now. The study's findings reveal that by accelerating the deployment of renewable energy and scaling up clean technologies—including nuclear power where needed—we can create a power grid that not only meets our energy demands but also delivers substantial climate and health benefits. The costs of inaction are too great to ignore. By taking bold action today, we can avoid billions in climate-related damage and save thousands of lives through cleaner air and a healthier environment.

References

1. U.S. Energy Information Administration. (April 2022). Monthly Energy Review, Table 1.3 and 10.1, preliminary data. - U.S. Energy Information Administration Search Results. (n.d.). https://search.usa.gov/search?utf8=%E2%9C%93&affiliate=eia.doe.gov&query=1.%09U.S.+Energy+Information+Administration.+%28April+2022%29.+Monthly+Ener-

gy+Review%2C+Table+1.3+and+10.1%2C+preliminary+data.&search=

2. *Carbon Capture, utilisation and Storage - Energy System - IEA. (n.d.).* IEA. https://www.iea.org/energy-system/carbon-capture-utilisation-and-storage

3. *Direct Air Capture - Energy System - IEA. (n.d.).* IEA. https://www.iea.org/energy-system/carbon-capture-utilisation-and-storage/direct-air-capture

4. *Renewable energy.* (n.d.). https://education.nationalgeographic.org/resource/renewable-energy/

5. Ritchie, H., Roser, M., & Rosado, P. (2024, March 11). *Renewable energy.* Our World in Data. https://ourworldindata.org/renewable-energy

6. Parfit, M. (n.d.). Future Power: Where will the world get its next energy fix? *Environment.* https://www.nationalgeographic.com/environment/article/powering-the-future

7. *Wave and Tidal Power Analysis - 1517 words | Report Example.* (2022, March 22). IvyPanda. https://ivypanda.com/essays/wave-and-tidal-power-analysis/

8. Torres, J. F., & Petrakopoulou, F. (2022). A closer look at the environmental impact of solar and wind energy. *Global Challenges, 6*(8). https://doi.org/10.1002/gch2.202200016

9. *Renewable energy.* (n.d.). Union of Concerned Scientists. https://www.ucsusa.org/energy/renewable-energy

10. Bošnjaković, M., Santa, R., Crnac, Z., & Bošnjaković, T. (2023). Environmental impact of PV power systems. *Sustainability, 15*(15), 11888. https://doi.org/10.3390/su151511888

11. *Accelerating clean energy ambition.* (2023, November 16). Union of Concerned Scientists. https://www.ucsusa.org/resources/accelerating-clean-energy-ambition

12. *100% clean electricity by 2035 study.* (2022, November 7). Energy Analysis | NREL. https://www.nrel.gov/analysis/100-percent-clean-electricity-by-2035-study.html

Chapter 4

The Role of Nuclear Power in Combating Climate Change: An Overview

While the fossil fuels and traditional green movements face an inconvenient truth: nuclear power stands as the world's most abundant and highly scalable source of nearly carbon-free energy. This fact challenges the prevailing narrative around renewable energy sources and has prompted us to reconsider the potential of adding more nuclear power to our green energy mix in combating climate change. In the face of the growing threat of climate change, finding enough sustainable and low-carbon sources of energy has become an urgent priority. Here we'll explore how nuclear power can potentially combat global warming, providing a concise overview of key facts and expert opinions on the matter.

What is Nuclear Power and How Does It Produce Electricity?

Nuclear power harnesses energy from the controlled splitting of atoms, known as nuclear fission. This process releases a significant amount of heat, which is then used to generate electricity. Unlike fossil fuels, nuclear power does not emit greenhouse gases during electricity generation, making it a low-carbon energy source. In 1954, the Obninsk Nuclear Power Plant in the USSR became the world's first nuclear power plant to generate electricity for a power grid. Today, nuclear power plants operate in many countries—likely more than you might imagine. You've heard almost nothing about them in recent years because they have been operating, producing electricity—a lot of electricity—quietly and safely.

Nuclear power plants and fossil fuel plants both generate electricity by using heat to produce steam, which then drives turbines connected to generators. However, the methods they use to produce this heat are fundamentally different.

In nuclear power plants, heat is generated through nuclear fission, where uranium atoms are split to release energy. This occurs in a reactor core, where uranium fuel pellets are stacked in fuel rods and bundled into assemblies. The fission process is carefully controlled using control rods that absorb neutrons, regulating the reaction rate—a precise control mechanism unique to nuclear reactors.

Fossil fuel plants, on the other hand, produce heat by burning coal, oil, or natural gas. This combustion process is simpler but releases significant amounts of carbon dioxide (CO_2), methane (CH_4), and other pollutants into the atmosphere. The heat transfer process also differs between nuclear and fossil fuel plants. Nuclear plants often use pressurized water reactors (PWRs) or boiling water reactors (BWRs). In PWRs, water under high pressure transfers heat to a secondary water system via steam generators, while in BWRs, water boils directly in the reactor vessel. Fossil fuel plants typically produce steam directly by burning fuel.

Perhaps the most significant difference lies in their environmental impact. Nuclear power plants produce electricity without emitting greenhouse gases during operation, making them a low-carbon energy source. In contrast, fossil fuel plants are major contributors to global CO_2 emissions and air pollution, driving global warming and harming public health. While both types of plants ultimately use steam to generate electricity, the unique aspects of nuclear fission, fuel type, reactor control, and environmental impact set nuclear power apart from fossil fuel generation.

U.S. Operating Commercial Nuclear Power Reactors

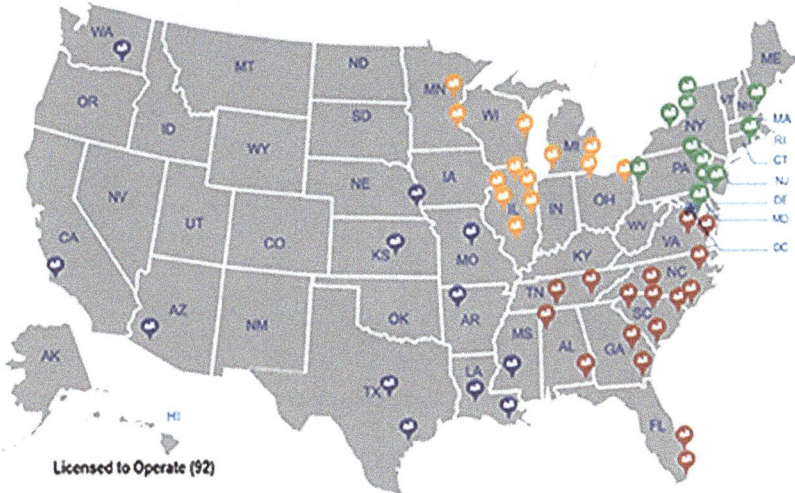

Licensed to Operate (92)

Source: U.S. Nuclear Regulatory Commission - As of February 2023

Global Nuclear Power

Nuclear power plays a significant role in the global energy landscape, with approximately 440 commercial nuclear reactors operating worldwide, and an additional 61 reactors under construction. These facilities are spread across about 32 countries, primarily concentrated in Europe, North America, and East Asia. The United States leads with 94 operating reactors as of 2024, while France stands out for generating around 70% of its electricity from nuclear power. China has emerged as the second-largest producer of nuclear electricity globally, with Russia also maintaining substantial nuclear power capacity.

Despite being present in only a fraction of the world's countries, nuclear power has a disproportionate impact on global electricity production, generating approximately 10% of the world's electricity in 2022. Nuclear energy is particularly notable for its role in low-carbon electricity generation, providing about one-quarter of the world's low-carbon power. In fact, it ranks as the second-largest source of low-carbon electricity, accounting for 26% of the total in 2020.

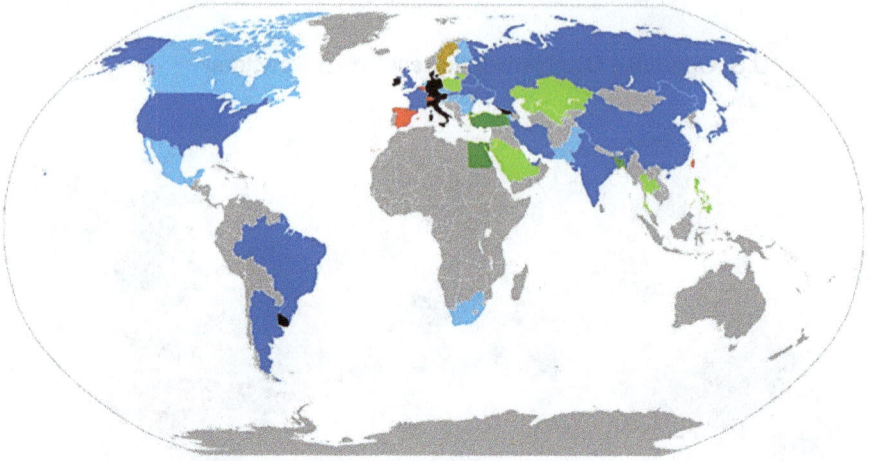

World Nuclear Power Generating Countries (in Color)

Global status of nuclear deployment as of April 2023 ▪ Operating reactors, building new reactors ▪ Operating reactors, planning new build ▪ No reactors, building new reactors ▪ No reactors, planning new build ▪ Operating reactors, stable ▪ Operating but may phase-out ▪ Civil nuclear power is illegal ▪ No reactors

The influence of nuclear power extends beyond the borders of countries with reactors. Many nations, especially in Europe, benefit from nuclear-generated electricity through regional transmission grids, highlighting the interconnected nature of modern energy systems. As the world grapples with the challenges of climate change and the need for clean energy sources, nuclear power will continue to be a significant contributor to the global energy mix, offering a steady supply of low-carbon electricity.

Key Facts on Nuclear Power and Climate Change:

Carbon Emissions Reduction

Nuclear power plants do not produce carbon dioxide or other greenhouse gases during electricity generation. According to the International Atomic Energy Agency (IAEA), nuclear power's lifecycle emissions are comparable to those of renewable energy sources like wind and solar. This characteristic makes nuclear power a highly viable option for reducing carbon emissions and combating global warming.

High Energy Density

Nuclear power boasts a high energy density, meaning a very small amount of nuclear fuel can produce a large amount of electricity. This efficiency is crucial for meeting the growing global energy demand while minimizing greenhouse gas emissions.

Baseload Power

Nuclear power plants provide a consistent and reliable source of electricity, known as baseload power. Unlike solar or wind power, which are intermittent, nuclear power can operate continuously, ensuring a stable electricity supply to the grid and reducing dependence on fossil fuels.

The following graph illustrates which countries have already embraced nuclear and renewables for a significant portion of their electricity production and those which still have a long way to go.

Electricity consumption from fossil fuels, nuclear and renewables, 2022

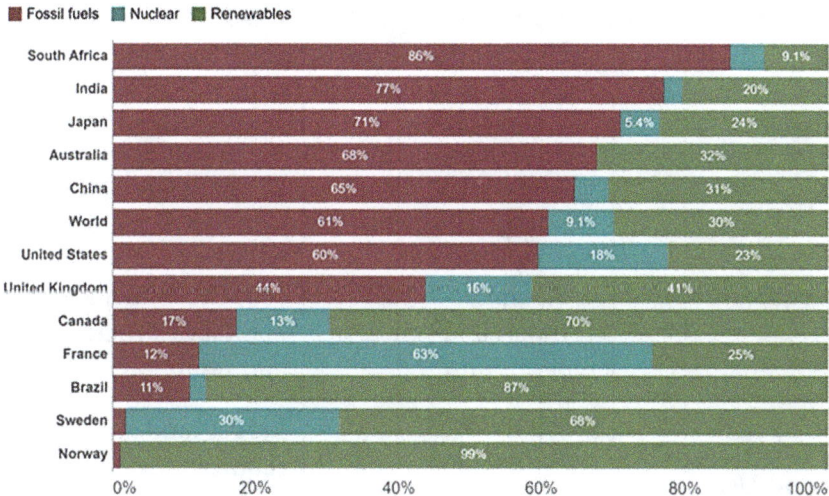

Source: Ember's Yearly Electricity Data; Ember's European Electricity Review; Energy Institute Statistical Review of World Energy
OurWorldInData.org/energy • CC BY

In the fight against global warming, nuclear power stands as a crucial tool. Its ability to generate electricity without emitting greenhouse gases, combined with its high energy density and baseload power characteristics, makes it an attractive option for reducing carbon emissions.

While concerns about safety and waste management are real, ongoing advancements in nuclear technology are addressing these challenges. With careful planning and regulation, nuclear power can significantly contribute to combating global warming and transitioning to a sustainable energy future.

When we closely examine nuclear power, we find it holds immense potential. Uranium, the primary fuel for nuclear reactors, is abundant and found in significant quantities across the globe, making it a viable long-term energy solution. Unlike other non-renewable energy sources, such as fossil fuels, nuclear power benefits from highly abundant sources that are not subject to depletion in the foreseeable future.

Nuclear vs. Fossil Fuels

Let's take a moment to compare the mining, land use, environmental impact, and resource availability between nuclear and fossil fuel electricity production. The following table clearly outlines these key differences.

Key Differences Between Nuclear & Fossil Fuel Electricity Generation

Aspect	Nuclear Power	Fossil Fuels
Mining Requirements	Uranium mining (underground and open-pit). Lower environmental impact compared to other types of mining.	Coal, oil, and gas extraction. Significant environmental consequences, including habitat destruction and pollution.
Land Use Requirements	Relatively small land footprint. Additional land needed for long-term waste storage.	Larger land areas required. Extensive space for fuel storage, ash ponds, and cooling infrastructure.

Aspect	Nuclear Power	Fossil Fuels
Environmental Impact	Zero greenhouse gas emissions during operation. Environmental concerns from uranium mining and waste storage.	Major contributor to greenhouse gas emissions and climate change. Air and water pollution, habitat destruction.
Resource Availability	Uranium reserves sufficient for foreseeable future. Less abundant than fossil fuels but significant global reserves.	Finite and non-renewable resources. Becoming more challenging and expensive to access remaining reserves.

It is important to keep in mind that this comparison is complex, and there are additional factors to consider, such as cost, safety, and public perception.

Scalability is another key attribute of nuclear power. While renewable energy sources like wind and solar have made significant strides, they still face limitations in terms of scalability. Nuclear power, on the other hand, can provide a substantial amount of energy consistently, making it well-suited to meet the growing global energy demand. The ability of nuclear power plants to generate baseload power, operating continuously and reliably, enhances its scalability and positions it as a dependable energy source.

Safety is of utmost importance in today's nuclear industry. Modern nuclear power plants are designed with multiple layers of safety measures to prevent accidents. The International Atomic Energy Agency (IAEA) and national regulatory bodies enforce strict safety standards to protect people and the environment. Additionally, advancements in waste management techniques have made nuclear power an even more viable option. Technologies like reprocessing and advanced storage methods have improved the handling and disposal of nuclear waste.

The future of nuclear power holds great promise with the development of advanced reactor technologies. These next-generation reactors aim to enhance safety, increase efficiency, and reduce waste production. For example, small modular reactors (SMRs) are being developed to provide flexible and scalable nuclear power solutions. Furthermore, research is underway on advanced fuel cycles and fusion energy, which have the potential to revolutionize the nuclear industry and further reduce environmental impacts.

However, it's important to acknowledge that nuclear power is not without its challenges. Safety, waste management, and public perception are areas that require careful consideration and continuous improvement, and I will be discussing them later in this book. Advancements in technology and strong regulatory frameworks have made nuclear power safer and more efficient than ever before. Public understanding and support are crucial in shaping the future of nuclear energy.

Expert Opinions

According to many experts, nuclear power can play a significant role in our transition to a clean energy future and combat global warming:

- **The International Atomic Energy Agency (IAEA)** states that nuclear power is a "dispatchable low-carbon source of electricity" and can contribute to decarbonization efforts.

- **A 2013 article by David Biello** argues that nuclear power is one of the few technologies that can quickly combat climate change, highlighting its importance in reducing carbon emissions and meeting our increasing energy demand.

- **James Hansen, a renowned climate scientist,** argues that nuclear power is essential for combating climate change. He notes that "the speediest drop in greenhouse gas pollution on record occurred in France in the 1970s and '80s, when that country transitioned from burning fossil fuels to nuclear fission for electricity."

- **Jeffrey Sachs, an economist and director of the Earth Institute at Columbia University,** supports the importance of nuclear energy in addressing climate change. He states, "On a global scale, it's hard to see how we could conceivably accomplish this [reducing global warming pollution] without nuclear."

- **The Nuclear Energy Institute (NEI)** emphasizes that "nuclear energy provides more than half of America's carbon-free electricity" and that "every year, nuclear-generated electricity saves our atmosphere from more than 470 million metric tons of carbon dioxide emissions that would otherwise come from fossil fuels."

- **The World Nuclear Association** highlights that nuclear power prevents the release of "2 billion tons of CO_2 each year across the world, equivalent to the production of 400 million cars."

- **Jean-Marc Jancovici, a recognized expert on energy and climate,** emphasizes "the urgent need to transition to low-carbon forms of energy" and considers nuclear power a key part of this transition.

Conclusion

Nuclear power has a significant role to play in combating climate change. Its low carbon footprint, reliability, and potential for advanced technologies make it an important component of a sustainable energy mix.

Public perception and policy are crucial in shaping the future of nuclear power. Policymakers must engage in open and transparent discussions about the benefits, risks, and challenges associated with nuclear energy. They need to create a supportive framework that incentivizes investment in nuclear power while ensuring safety in plant operations and waste management practices. Public awareness campaigns and education about nuclear energy can help dispel misconceptions and foster informed decision-making. By embracing nuclear power alongside other renewable energy sources, we can work toward a cleaner and more sustainable future.

There is a clear need for new clean energy generating capacity around the world, both to replace aging fossil fuel units, especially coal-fired plants, which emit greenhouse gases in large amounts, and to meet the increasing demand for electricity. In 2021, 61% of electricity was generated from the burning of fossil fuels. Despite the strong support for, and growth in, intermittent renewable electricity sources in recent years, the fossil fuel contribution to power generation has not changed significantly in the last 15 years (66.5% in 2005).

References

1. *Map of power reactor sites.* (2023, February). NRC Web. https://www.nrc.gov/reactors/operating/map-power-reactors. html
2. Wikipedia contributors. (2024, July 3). *Nuclear power by country*. Wikipedia. https://en.wikipedia.org/wiki/Nuclear_ power_by_country
3. *Nuclear power in the world today - World Nuclear Association.* (n.d.). https://world-nuclear.org/information-library/current-and-future-generation/nuclear-power-in-the-world-today

4. *Nuclear power in a clean energy system – Analysis - IEA.* (2019, May 1). IEA. https://www.iea.org/reports/nuclear-power-in-a-clean-energy-system

5. *Nuclear - IEA.* (2023, July 11). IEA. https://www.iea.org/energy-system/electricity/nuclear-power

6. Biello, D. (2013, December 12). *How nuclear power can stop global warming. Scientific American.* https://www.scientificamerican.com/article/how-nuclear-power-can-stop-global-warming/

7. *7 good reasons for turning to nuclear energy to combat global warming | Orano.* (n.d.). orano.group. https://www.orano.group/en/unpacking-nuclear/7-good-reasons-for-turning-to-nuclear-power-to-combat-global-warming

8. *Climate.* (n.d.). Nuclear Energy Institute. https://www.nei.org/advantages/climate

Chapter 5

Uranium: The Element Powering Our Future

In this chapter, we dive into the world of uranium, the primary fuel used in nuclear power plants. We'll explore its properties, its role in nuclear reactions, and its potential as a green solution for combating global warming. By understanding the basics of uranium, we can better appreciate the immense power and possibilities it holds for our energy future.

Uranium, a chemical element with the symbol U and atomic number 92, holds a fascinating history. Discovered in 1789 by the German chemist Martin Heinrich Klaproth, uranium has played a crucial role in various fields, including medicine and industry. For our purposes, it is most notable for its pivotal role in nuclear power generation.

This dense, silvery-white metal is slightly radioactive and naturally occurring in the Earth's crust, primarily found in the form of uranium ore. Uranium is a heavy metal with unique properties that make it ideal for use as nuclear fuel. The most common isotope of uranium is uranium-238, which accounts for over 99% of natural uranium. Two other isotopes, uranium-235 and uranium-234 are also present in the ore but in much smaller quantities.

U-238 has a very long half-life of over 4.4 billion years and undergoes alpha decay, emitting alpha particles. It is the parent isotope in a decay series that ends with the stable isotope lead-206. Due to its natural abundance and long half-life, U-238 significantly contributes to the background radiation on Earth. Although it is not fissile and cannot sustain a chain reaction with thermal neutrons, it is fissionable by fast neutrons and plays a crucial role in producing plutonium-239 in nuclear reactors, with applications in both military and civilian sectors, such as in depleted uranium.

One remarkable property of uranium-235 is its ability to undergo nuclear fission—a process where the nucleus of an atom splits into smaller parts, releasing an enormous amount of energy. This characteristic makes uranium an ideal fuel for nuclear reactors.

Uranium's significance extends beyond its basic chemical properties. It is a cornerstone in the quest for sustainable energy solutions. Unlike fossil fuels, uranium does not emit carbon dioxide during the energy production process, making it a potent ally in the fight against global warming. With advancements in nuclear technology, modern reactors are designed to maximize safety and efficiency, addressing many public concerns about nuclear energy.

Australia holds the largest known uranium reserves in the world, with about 28% of the global recoverable uranium resources. These reserves are significantly larger than those of the next closest countries, Kazakhstan and Canada. Despite having the largest reserves, Australia does not use nuclear power domestically and primarily exports its uranium.

In the context of combatting climate change, the role of uranium cannot be overstated. As we seek to reduce our reliance on carbon-intensive energy sources, uranium presents a viable, large-scale alternative. The element's capacity to generate vast amounts of clean energy makes it indispensable in our journey toward a sustainable and green future. By understanding uranium and its potential, we can make informed decisions about our energy policies and investments, ultimately contributing to a more stable and environmentally friendly world.

Nuclear Fission

Nuclear fission lies at the heart of nuclear power generation. Let's explore this fascinating process and see how nuclear reactors harness the power of uranium through nuclear fission.

Nuclear fission is a reaction in which the nucleus of a uranium atom is split into two smaller nuclei when bombarded by a neutron. This occurs in a controlled manner within the core of a nuclear reactor. Uranium-235 is particularly important in this process because it is fissile, meaning it can sustain a chain reaction of fission.

When a uranium-235 nucleus absorbs a neutron, it becomes highly unstable and splits into two smaller nuclei, along with the release of additional neutrons and a large amount of energy. These newly released neutrons can then collide with other uranium-235 nuclei, triggering a chain reaction. This chain reaction results in a continuous release of energy in the form of heat.

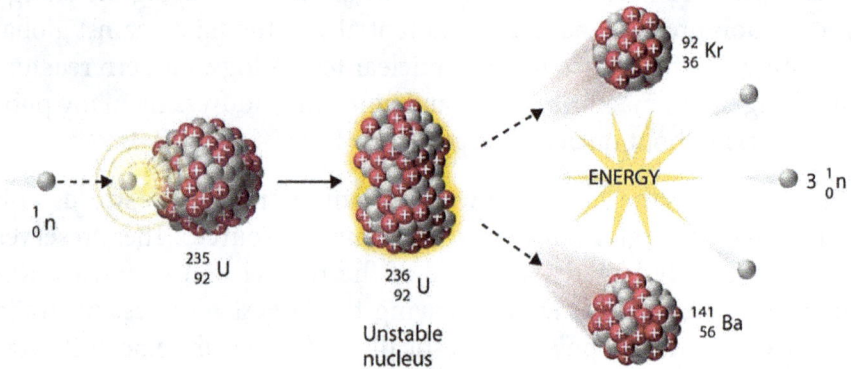

Basics of Nuclear Fission   *nuclear-power.com*

This energy release is described by Einstein's equation, $E=mc^2$, one of the most famous equations in physics. It relates energy (E) to mass (m) and the speed of light (c). The equation shows that the energy of an object is equal to its mass multiplied by the square of the speed of light. This concept revolutionized our understanding of the relationship between matter and energy, showing that mass can be converted into energy and vice versa. It explains the basis of nuclear energy, where the fission of uranium-235 leads to particles of less mass, and the remaining mass is converted into a large amount of energy.

Before uranium can be used as fuel in most reactors, it must undergo processes called enrichment and fuel fabrication. Since the concentration of U-235 in natural uranium is relatively low, typically around

0.7%, enrichment is employed to increase the U-235 content to around 3-5%, which is required for most reactors.

Enrichment is achieved through various methods, such as centrifugation or gaseous diffusion. Once enriched, the uranium is converted into a suitable form for fuel fabrication, typically as uranium dioxide pellets, which are stacked into fuel rods.

Understanding Nuclear Reactor Behavior

In nuclear reactors, reactivity measures how much the reactor's behavior deviates from a stable, self-sustaining state known as criticality. When a reactor is critical, the nuclear fission chain reaction continues at a constant rate, meaning the reactor's power output remains steady.

- Positive reactivity means the reactor is becoming supercritical, where the power output increases.

- Negative reactivity means the reactor is becoming subcritical, where the power output decreases.

- Zero reactivity means the reactor is critical, maintaining a stable power level.

In a critical state:

- Each fission event leads to exactly one new fission event, keeping the number of neutrons constant over time.

- The reactor's power output is constant, with no increase or decrease in the rate of fission reactions.

- The production of neutrons from fission perfectly balances the loss of neutrons due to absorption and leakage.

Operators aim to keep the reactor in this critical state during normal power generation by carefully managing factors like fuel composition and control rod positions.

It's important to understand that "critical" in this context does not imply danger; rather, it describes the desired, controlled operating condition of a nuclear power reactor.

The Diverse World of Nuclear Power Plants

Nuclear power plants have been a cornerstone of electricity generation worldwide for several decades. These engineering marvels harness the power of nuclear fission to produce clean, reliable energy. In this section, we'll take a closer look at the different types of nuclear power plants currently operating around the world, highlighting their unique features and contributions to the global energy landscape.

Pressurized Water Reactor (PWR)

The Pressurized Water Reactor is the most common type of nuclear power plant in operation today, particularly in the United States. PWRs use water as both a coolant and a moderator. In this context, a moderator is a substance that slows down the fast-moving neutrons produced during nuclear fission, allowing for a sustained chain reaction. The water in the reactor core absorbs the heat generated by nuclear fission and transfers it to a secondary water loop via a steam generator. PWRs are renowned for their high thermal efficiency, robust safety features, and proven track record. Notable examples of PWRs include the Palo Verde Nuclear Generating Station in Arizona, USA, and the Gravelines Nuclear Power Plant in France.

Boiling Water Reactor (BWR)

The Boiling Water Reactor also uses water as both a coolant and a moderator. However, unlike PWRs, the water in BWRs boils directly in the reactor core, producing steam that directly drives the turbine generator. This design simplifies the process and eliminates the need for a separate steam generator. BWRs are recognized for their simplicity and efficiency. The Kashiwazaki-Kariwa Nuclear Power Plant in Japan and the Dresden Generating Station in the United States are prominent examples of BWRs.

Heavy Water Reactor (HWR)

The Heavy Water Reactor uses heavy water (deuterium oxide) as a moderator and coolant. Heavy water slows down neutrons more effectively than regular water, allowing the use of natural uranium as fuel. HWRs are known for their flexibility in fuel selection and efficient use of resources. The Point Lepreau Nuclear Generating Station in Canada and the Cernavodă Nuclear Power Plant in Romania are key examples of HWRs.

Let me clarify a bit here. Heavy water contains a different form of the hydrogen atom, called deuterium, which has a proton and a neutron. In heavy water, the hydrogen atoms are replaced with this heavier form, making the water molecule slightly denser than regular water, hence the name "heavy water." This difference gives heavy water unique properties that make it valuable in certain scientific and industrial applications, especially in nuclear reactors.

In normal water, the percentage of heavy water is very low—approximately 0.015% of the total water content. While it is not inherently dangerous or harmful to humans, heavy water must be concentrated for use in reactors through a complex and energy-intensive process called isotopic exchange. Due to its density, heavy water is more effective at slowing down neutrons, allowing for more controlled and efficient nuclear reactions in HWRs.

Gas-Cooled Reactor (GCR)

The Gas-Cooled Reactor uses carbon dioxide (CO_2) gas as both the coolant and the moderator. The CO_2 circulates through the reactor core, absorbing heat from the fuel assemblies. GCRs are known for their high-temperature operation and potential for cogeneration, where excess heat can be used for industrial processes or district heating. The Advanced Gas-Cooled Reactor in the United Kingdom is a well-known example of a GCR.

Fast Breeder Reactor (FBR)

The Fast Breeder Reactor (FBR) is designed to produce more fissile material, such as plutonium, than it consumes. These reactors use fast neutrons to sustain the nuclear chain reaction and can employ various coolants, such as liquid sodium or lead. FBRs hold the potential to significantly increase the utilization of nuclear fuel resources. Notable examples include the Monju reactor in Japan and the BN-800 reactor in Russia.

Nuclear Power Plant Examples

Pressurized Water Reactors: The Most Common Type of Nuclear Power Plant in Operation Today

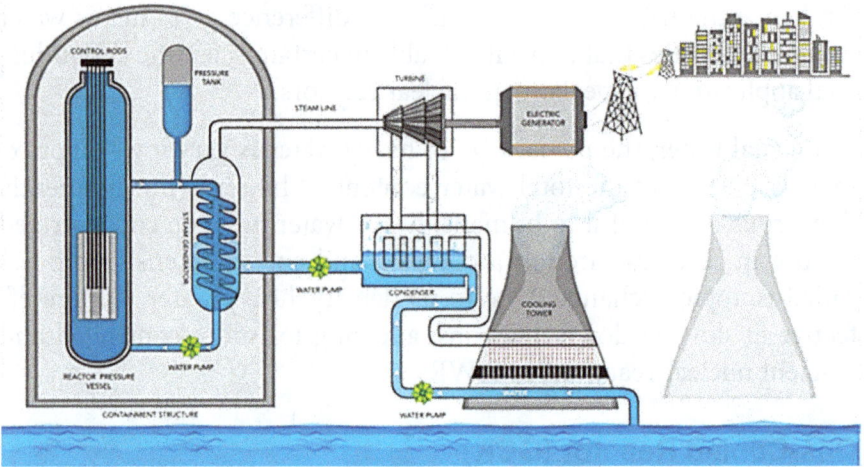

PRESSURIZED WATER REACTOR (PWR)

A Pressurized Water Reactor (PWR) is a sophisticated system that uses water as both the coolant and moderator to safely and efficiently generate electricity. Here's a streamlined explanation of how a PWR operates:

- **Reactor Core:** The core of a PWR is its heart, containing fuel assemblies filled with uranium fuel pellets. This is where nuclear fission reactions take place, all enclosed within a robust pressure vessel.

- **Neutron Moderation:** The water in the reactor acts as a moderator, slowing down the fast-moving neutrons released during fission. This moderation is crucial for sustaining a controlled chain reaction, with the level of moderation carefully adjusted to maintain optimal reactor performance.

- **Heat Generation:** As uranium fuel undergoes fission, it releases a tremendous amount of heat, which is absorbed by the surrounding water coolant.

- **Primary Coolant Circuit:** The heated water within the reactor core, known as the primary coolant, circulates through the core, absorbing heat generated by the nuclear reactions in the fuel assemblies.

- **Steam Generator:** The primary coolant transfers its heat to a secondary water loop through a heat exchanger called the steam generator. This secondary loop is isolated from the reactor and contains non-radioactive water, ensuring an additional layer of safety.

- **Steam Production:** The heat from the primary coolant boils the water in the secondary loop, producing high-pressure steam.

- **Steam Turbine:** The high-pressure steam is directed into a turbine, causing it to spin. This spinning turbine is connected to a generator, converting kinetic energy into electrical energy, which is then transmitted through power lines to homes, businesses, and industries, ensuring a reliable and continuous electricity supply.

- **Condenser and Cooling:** After passing through the turbine, the steam enters a condenser, where it is cooled and condensed back into liquid form using cold water from a nearby source. This non-radioactive water is then safely discharged back into the environment.

- **Feedwater:** The condensed water, known as feedwater, is pumped back into the steam generator to be heated and converted into steam again, maintaining the continuous cycle of power generation.

- **Control and Safety Systems:** PWRs are equipped with various control and safety systems that continuously monitor and regulate the nuclear reactions, ensuring reactor stability and enabling a safe shutdown in case of emergencies.

Overall, a PWR operates by harnessing the energy from nuclear fission to generate heat, which is then used to produce steam that drives a turbine and generator. The closed-loop system, with separate primary and secondary circuits, enhances both the safety and efficiency of the nuclear power plant.

CANDU Reactors: A Versatile and Safe Nuclear Power Solution

CANDU, an acronym for "CANada Deuterium Uranium," is a unique type of heavy water nuclear power reactor (HWR) developed in Canada. Like other reactors, it generates steam to drive a turbine and produce electricity, but CANDU reactors are distinguished by their distinct features and capabilities, making them a significant part of the global nuclear power landscape. Let's explore the key aspects of CANDU reactors, including their fuel flexibility, safety features, and on-power refueling capabilities.

- **Fuel Flexibility and Cost-Effectiveness:** One of the standout features of CANDU reactors is their ability to use natural uranium as fuel, setting them apart from other reactor types that require enriched uranium. This fuel flexibility makes CANDU reactors especially advantageous for countries with abundant natural uranium resources. By directly utilizing unenriched uranium, CANDU reactors offer a cost-effective option for generating nuclear power.

- **Safety Features Enhancing Stability:** CANDU reactors possess an inherent safety feature known as a "negative void coefficient." This means that as the coolant temperature rises, the reactor's reactivity decreases, significantly reducing the risk of a runaway chain reaction. This safety feature enhances the stability and reliability of CANDU reactors, making them a secure choice for nuclear power generation.

- **On-Power Refueling for Greater Flexibility:** Another notable capability of CANDU reactors is their on-power refueling. Unlike traditional reactors that require shutdowns for refueling, CANDU reactors can replace fuel bundles while still in operation. This unique feature provides greater flexibility in maintenance and fuel management, minimizing downtime and maximizing power output.

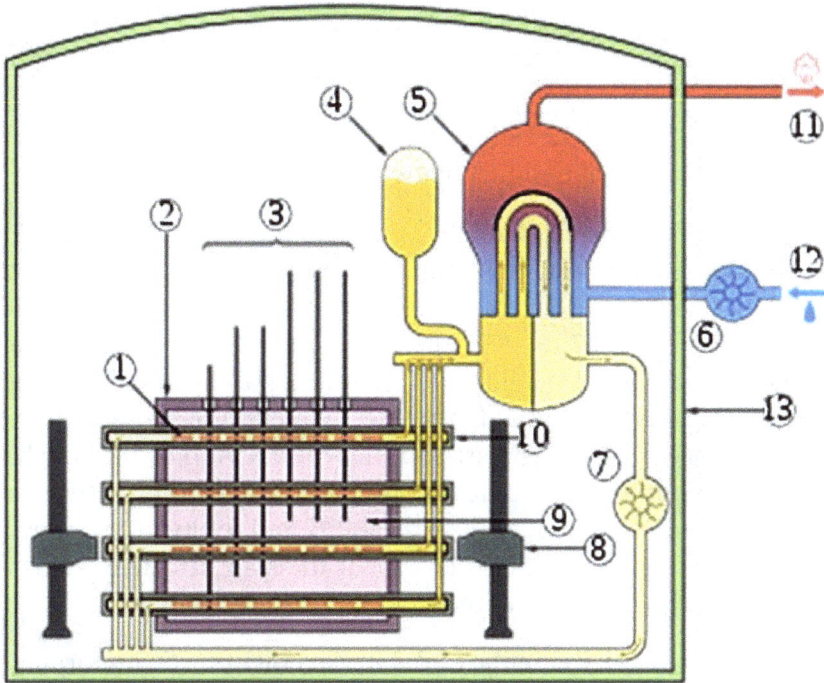

Schematic Diagram of a CANDU Reactor:

Hot and cold sides of the primary heavy-water loop; hot and cold sides of secondary light-water loop; and cool heavy water moderator in the calandria, along with partially inserted adjuster rods (as CANDU control rods are known).

1. Fuel bundle
2. Calandria (reactor core)
3. Adjuster rods
4. Pressurizer
5. Steam generator
6. Light-water pump
7. Heavy-water pump
8. Fueling machines
9. Heavy-water moderator

10. Pressure tube

11. Steam going to steam turbine

12. Cold water returning from turbine

13. Containment building made of reinforced concrete

Proven Success and Global Deployment: The deployment of CANDU reactors began with the first CANDU power reactor, the Nuclear Power Demonstration (NPD) reactor, located in Ontario, Canada. Operating from 1962 to 1987, the NPD reactor played a crucial role in demonstrating the viability and capabilities of the CANDU design. This success paved the way for the deployment of larger-scale CANDU reactors in Canada and around the world. CANDU reactors have been successfully deployed in various countries, including Argentina, Romania, and South Korea. Their reliability and ability to generate significant amounts of electricity have made them an attractive option for countries seeking to expand their nuclear power capabilities.

Continuous Advancements and Improvements: It's important to note that the CANDU design has evolved over time, with successive generations and iterations of the technology. Each generation incorporates improvements in safety, efficiency, and fuel utilization. These advancements ensure that CANDU reactors remain at the forefront of nuclear power technology, consistently meeting the ever-increasing demands for safe and sustainable energy.

The CANDU reactor design has proven to be a versatile and safe solution for nuclear power generation. Its unique features—such as the use of natural uranium, on-power refueling, and inherent safety mechanisms—have made it a reliable choice for countries looking to harness nuclear energy. With a successful track record and continuous advancements, CANDU reactors play a significant role in meeting the world's growing energy needs and will continue to do so in the future.

Other Nuclear Reactors

Beyond commercial nuclear power plants, about 220 research reactors are operating in over 50 countries, with more under construction. These reactors, aside from their use in research and training, produce essential medical and industrial isotopes.

Nuclear reactors also play a crucial role in marine propulsion, predominantly within the major navies. For five decades, they have powered submarines and large surface vessels. Over 160 ships, mainly submarines, are propelled by approximately 200 nuclear reactors, accumulating over 13,000 reactor years of experience. Both Russia and the USA have decommissioned many of their Cold War-era nuclear submarines, but Russia continues to operate a fleet of large nuclear-powered icebreakers, with additional units under construction. Additionally, Russia has connected a floating nuclear power plant with two 32 MWe reactors to the grid in the remote Arctic. These reactors are adapted from those used in icebreakers, showcasing the versatility and adaptability of nuclear technology.

Conclusion

One of the most remarkable aspects of nuclear fission is the immense amount of energy it releases. The energy density of nuclear fuel, such as uranium, is millions of times higher than that of fossil fuels. This means that a very small amount of uranium can produce a vast amount of energy. Understanding uranium—the primary fuel for nuclear reactors—is vital for grasping the full potential of nuclear power. With its unique properties, uranium enables the process of nuclear fission, generating substantial amounts of clean energy. Rigorous safety measures and waste management protocols ensure the responsible use and disposal of nuclear fuel. By embracing and expanding our knowledge of uranium, we can fully appreciate its role in generating sustainable and low-carbon energy, ultimately contributing to the fight against global warming.

By harnessing the power of nuclear fission, nuclear reactors provide a reliable and abundant source of electricity that is virtually free of greenhouse gas emissions. This makes nuclear power a valuable component of the energy mix in combating climate change and meeting the growing energy demands of our modern world.

References

1. *Uranium-235 (U-235) and uranium-238 (U-238).* (2024, April 17). Radiation Emergencies. https://www.cdc.gov/radiation-emergencies/hcp/isotopes/uranium-235-238.html
2. *Nuclear power in the world today - World Nuclear Association.* (2024, May 7). https://world-nuclear.org/information-library/current-and-future-generation/nuclear-power-in-the-world-today
3. Wikipedia contributors. (2024, July 3). *Nuclear power by country.* Wikipedia. https://en.wikipedia.org/wiki/Nuclear_power_by_country
4. Naval Supply Systems Command. (n.d.). *Naval Nuclear Propulsion Overview.* https://www.navsup.navy.mil/Viper-Home/NNPO/

Chapter 6

Safety First: Examining Safety Measures in Nuclear Power Plants

Let's review these significant nuclear accidents, assess their human costs, and contrast them with the often-overlooked impacts of fossil fuel-generated electricity.

1. Three Mile Island (1979):

The Three Mile Island nuclear accident was a partial meltdown of the reactor core at the Three Mile Island Nuclear Generating Station near Harrisburg, Pennsylvania. It remains the most significant nuclear accident in U.S. history. This incident was caused by a combination of equipment malfunctions, design flaws, and human errors. A series of failures led to a loss of coolant, resulting in a partial meltdown of the reactor core. Despite these failures, the containment structure effectively prevented a complete release of radioactive materials, although a small amount of radioactive gas and iodine was released into the environment.

In the immediate aftermath, widespread public fear and uncertainty ensued due to concerns about the extent of the release and potential health effects. As a precaution, approximately 140,000 people, primarily pregnant women and young children, were temporarily evacuated. However, studies conducted by the National Cancer Institute (NCI) in the years following the accident found no evidence of long-term health impacts or increased cancer rates among the general population exposed to radiation. The NCI concluded that the radiation doses received by the public were well below harmful levels.

Environmentally, the immediate effects included the contamination of water and soil in the vicinity of the plant. However, subsequent studies revealed that the environmental impact was relatively small and localized. The area around Three Mile Island has been monitored for long-term effects, with no significant impacts observed.

Interestingly, the other reactor at the site continued operation for 40 years before shutting down in 2019, generating over 800 megawatts of carbon-free electricity and providing employment for 675 people at its peak.

2. Chernobyl (1986):

The Chernobyl disaster in Ukraine is widely regarded as the most severe nuclear power accident in history. It was primarily caused by a combination of design flaws in the reactor and operator errors during a safety test. The Chernobyl reactor design had inherent safety issues, including increased core reactivity as coolant water turned into steam, making the reactor unstable at low power levels.

During the safety test, operators attempted to simulate a power outage and the subsequent activation of emergency systems. However, due to a series of mistakes and misinterpretations of the test procedure, reactor power levels dropped dangerously low, worsening instability. In an attempt to recover power levels, the operators inadvertently caused a rapid increase in reactivity, leading to a power surge and explosion.

The explosion and ensuing fire released a massive amount of radioactive material into the atmosphere, affecting vast areas of Ukraine, Belarus, Russia, and parts of Western Europe. The accident was a result of technical and human factors, including design flaws, inadequate operator training, and a lack of safety culture. According to the United Nations Scientific Committee on the Effects of Atomic Radiation (UNSCEAR), the immediate death toll was 31, including plant workers and firefighters. The long-term effects are more difficult to quantify, but it is estimated that the accident may have led to several thousand additional cancer-related deaths.

3. Fukushima (2011):

The Fukushima accident in Japan was triggered by a massive earthquake (Magnitude 9) and subsequent tsunami, resulting in a nuclear meltdown and the release of radioactive materials. The primary cause of direct deaths related to the accident was the evacuation process, which posed significant challenges, particularly for elderly and vulnerable individuals. However, the World Health Organization (WHO) reported that radiation exposure levels were too low to cause a noticeable increase in cancer rates among the general public. There was one confirmed death from radiation four years later, attributed to lung cancer. It is important to note that the earthquake and tsunami caused over 15,000 casualties unrelated to the nuclear reactors' destruction at Fukushima.

Lessons Learned and Safety Advances

The lessons from these past incidents have driven significant advancements in safety practices and regulations. Today's nuclear power plants are equipped with multiple layers of safety systems to prevent accidents and mitigate potential risks.

Comparing Nuclear Power Accidents with Fossil Fuel Impacts

While past nuclear accidents have captured substantial attention due to their catastrophic nature, it is essential to consider the broader context of energy-related risks. Fossil fuel-generated electricity also has significant health and environmental impacts. Air pollution from burning fossil fuels contributes to respiratory diseases, cardiovascular problems, and premature deaths. The World Health Organization estimates that outdoor air pollution alone leads to around 4.2 million premature deaths annually.

Death rates per unit of electricity production

Death rates are measured based on deaths from accidents and air pollution per terawatt-hour (TWh) of electricity.

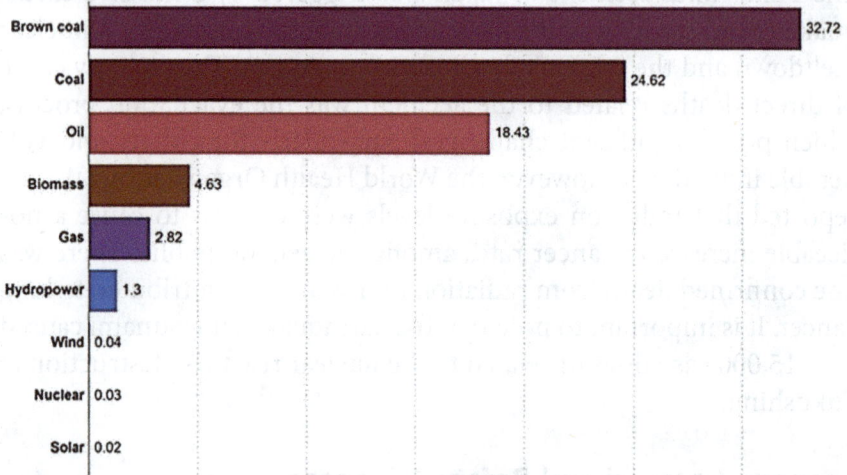

Our World in Data

Brown coal	32.72
Coal	24.62
Oil	18.43
Biomass	4.63
Gas	2.82
Hydropower	1.3
Wind	0.04
Nuclear	0.03
Solar	0.02

Source: Markandya & Wilkinson (2007); Sovacool et al. (2016); UNSCEAR (2008; & 2018)
OurWorldInData.org/energy • CC BY

From this graph, you can see that the death rates for wind, nuclear, and solar power production are very, very low, and only a tiny percentage of the death rates caused by electricity produced from coal and oil—death rates that we, as a society, have been accepting for many years.

Safety Measures at Nuclear Plants

Now, let's delve into the crucial topic of safety in nuclear power plants. Safety remains paramount in the nuclear industry, and modern reactors are equipped with comprehensive safety measures to ensure the protection of people and the environment. Join me as we explore the robust safety protocols and technologies that make nuclear power one of the safest forms of energy generation today.

Understanding Safety Culture

Safety culture is the bedrock of a nuclear power plant's operations. It encompasses the attitudes, behaviors, and practices that prioritize safety above all else. A strong safety culture fosters a proactive approach to identifying and addressing potential risks, ensuring continuous improvements, and learning from past experiences.

Various organizations, including the International Atomic Energy Agency (IAEA), provide guidelines and promote the development of a strong safety culture in the nuclear industry.

Defense in Depth

One of the fundamental principles of nuclear safety is "defense in depth." Modern nuclear power plants are designed with multiple layers of protection, each independently designed to prevent the release of radioactive material. This approach involves creating multiple barriers and redundant systems to prevent accidents and mitigate their consequences. These include physical barriers, such as containment structures, and engineered safety features like emergency cooling systems, automatic shutdown mechanisms, and robust emergency preparedness plans.

Physical Barriers

Nuclear power plants are constructed with physical barriers to contain radioactive material and prevent its release during normal operations and accidents. These barriers include thick concrete walls, containment structures, and reactor vessel designs that can withstand extreme conditions. They act as the first line of defense against the release of radiation.

Redundant Safety Systems

Nuclear power plants are outfitted with redundant safety systems to ensure that even if one system fails, backup systems are in place to maintain safety. These systems include emergency cooling systems, backup power supplies, and emergency shutdown mechanisms. They are rigorously tested and regularly maintained to ensure their reliability.

Rigorous Training and Maintenance

The nuclear industry places an unparalleled emphasis on rigorous training and maintenance programs. Plant operators undergo intensive training to ensure they can prevent and respond effectively to any potential safety concerns. Regular and thorough inspections and maintenance activities are conducted to promptly identify and address any issues. Furthermore, advanced technologies like computerized monitoring systems and predictive analytics facilitate the early detection of warning signs, enabling proactive maintenance.

Mitigating the Consequences of a Nuclear Accident

In the unlikely event of an incident, beyond the systems mentioned earlier to mitigate consequences, emergency preparedness and response plans are meticulously developed by plant operators in collaboration with federal, state, and local authorities to ensure effective coordination and communication. These comprehensive plans include evacuation procedures, medical response capabilities, and measures to prevent the release of radioactive materials. Periodic tests of these plans ensure that all parties with emergency responsibilities can work together effectively to protect the public and the environment should the need arise.

In the U.S., this planning and testing are overseen by the Nuclear Regulatory Commission (NRC) in coordination with the Federal Emergency Management Agency (FEMA). The two planning zones are:

- **Within a 10-mile Emergency Planning Zone (EPZ)**, immediate protective actions for the public may include instructions for sheltering in place or evacuation.

- **Within a 50-mile zone**, federal and state governments monitor and test food and water supplies that could potentially become contaminated and, if necessary, remove any found to be unsafe from public consumption.

The U.S. Nuclear Regulatory Commission

The Nuclear Regulatory Commission (NRC) is an independent federal agency that acts as a strong and effective regulator of commercial nuclear power plants in the United States. The NRC issues federal licenses to construct and operate nuclear power plants only after thorough safety, security, and environmental reviews.

The NRC not only oversees the construction of nuclear facilities but also monitors their operation through rigorous inspections and performance reviews.

The agency's mission is to ensure the adequate protection of public health and safety, promote common defense and security, and protect the environment. This mission is fulfilled by ensuring that each plant complies with the technical and administrative requirements established by the agency and adheres to the terms imposed by its facility license.

Enhancing Safety Through International Cooperation

The nuclear industry actively promotes international cooperation to enhance safety standards and knowledge sharing. Organizations like the World Association of Nuclear Operators (WANO) facilitate peer reviews and benchmarking exercises to identify best practices and areas for improvement. The sharing of data and experiences among countries with nuclear power programs fosters a collective learning process that benefits the global nuclear community.

Could a Nuclear Power Plant Ever Explode Like an Atomic Bomb?

Before leaving this topic of safety, it's important to address a common but unwarranted concern: Could a Nuclear Power Plant Ever Explode Like an Atomic Bomb? The short answer is a definite "no!" A nuclear power plant cannot and will not explode like an atomic bomb. Although both nuclear power plants and atomic bombs involve nuclear reactions, and uranium fuel is used in both, the processes and conditions are vastly different.

In a nuclear power plant, the nuclear reaction is carefully controlled to generate electricity. The fuel used, usually uranium or plutonium, is in the form of fuel rods and is enriched to a level suitable for power generation. The nuclear reaction in a power plant is sustained by maintaining a controlled chain reaction, with neutron moderation and cooling systems continuously managing the heat generated.

On the other hand, an atomic bomb, also known as a nuclear weapon, relies on an uncontrolled chain reaction. This means that the reaction escalates extremely rapidly, resulting in a massive release of energy in the form of an explosion. Atomic bombs use highly enriched uranium or plutonium and are engineered specifically to maximize destructive power.

The enrichment percentages of uranium used in nuclear power plants and those required to build an atomic bomb are fundamentally very different. In nuclear power plants, the uranium fuel is typically enriched to a level between 3% and 5% uranium-235 (U-235) isotopes. This level of enrichment is sufficient to sustain a controlled chain reaction for electricity generation. The majority of the uranium consists of the U-238 isotope, which is not fissile (cannot sustain a chain reaction) at the low enrichment levels used in power plants.

Conversely, the enrichment level required for building an atomic bomb is much higher. To create a nuclear weapon, uranium-235 or plutonium-239 (Pu-239) must be enriched to a level above 90%. Such a high enrichment level is necessary to achieve a supercritical mass and sustain an uncontrolled chain reaction, resulting in an incredibly powerful explosion.

Enriching uranium to such high levels is technically challenging, requiring advanced facilities and expertise. It involves techniques like gas centrifugation or gaseous diffusion to separate the U-235 isotope from the U-238 isotope. These enrichment methods are highly regulated and closely monitored to prevent the proliferation of nuclear weapons.

So, rest assured that the design and construction of nuclear power plants are fundamentally different from those of atomic bombs. And as I've discussed earlier, power plants have multiple layers of safety systems and controls to prevent uncontrolled chain reactions. These safety measures include control rods, which absorb neutrons and help regulate the rate of the reaction, as well as emergency cooling systems to prevent overheating. And the newer reactor designs minimize or even eliminate the possibility of overheating.

Let me remind you that nuclear accidents, like the Chernobyl disaster in 1986, can release large amounts of radioactive material and have severe consequences. However, such accidents are not, and cannot be, large explosions in the same sense as an atomic bomb.

Possible Terrorist Attacks and War

I'll start here with some general information and concerns, then move on to the current, real concerns caused by the war in Ukraine. In addition to the safety measures mentioned above, nuclear power plants implement stringent security measures to prevent unauthorized access and protect against potential terrorist attacks. These measures include physical barriers, surveillance systems, armed guards, and security protocols. Additionally, emergency response plans are in place to mitigate the consequences of an attack and protect the surrounding areas. We must acknowledge that the possibility of a terrorist attack on a nuclear power plant is a serious matter for security experts.

The major concern in the case of a terrorist attack on a nuclear power plant is that it could lead to the release of radioactive materials, resulting in severe consequences and damage to this critical infrastructure. These consequences may include radiation exposure, contamination of the surrounding area, and long-term health effects. The potential loss of electricity to the plant's customers for a significant period must also be considered. The severity of the impact would depend on various factors, such as the nature of the attack, the robustness of the plant's safety systems, and the effectiveness of emergency response measures. Thus, the effects of any attack would be unpredictable and depend on the specifics of the situation.

Risks associated with such attacks could include:

- **Sabotage**: Terrorists may attempt to sabotage critical systems within a nuclear power plant, such as disabling safety mechanisms or damaging control systems. This could potentially lead to a loss of control over the nuclear reaction or other hazardous situations.

- **Theft of nuclear material:** Nuclear power plants store radioactive materials that could be used to create a dirty bomb. A dirty bomb is a conventional explosive combined with radioactive material, which, when detonated, would spread radioactive particles over a wide area.

- **Disruption of cooling systems:** Most current and older nuclear power plants require continuous cooling to prevent overheating of the reactor core. Disrupting the cooling systems could lead to a reactor meltdown and the release of radioactive material.

It's important to note that specific security details and countermeasures are typically not disclosed publicly to avoid aiding potential attackers. However, regulatory bodies and government agencies responsible for nuclear safety continuously assess and update security measures to adapt to evolving threats.

Concerns Over Safety at Zaporizhzhia Nuclear Power Plant Amid Ongoing Conflict in Ukraine

The Zaporizhzhia Nuclear Power Plant (ZNPP), located in Ukraine, is Europe's largest nuclear power facility. With its six reactors boasting a total capacity of approximately 6,000 megawatts, it ranks among the most significant nuclear installations worldwide. The ongoing occupation of the plant by Russian forces since March 2022 has intensified international concern over its safety. Should a direct attack or bombing occur, the potential risks to the public and the environment could be catastrophic. Here, we examine the current situation at ZNPP, the control of the facility, and the potential consequences of a security breach.

Current Situation and Concerns

Radiation Levels: Despite the ongoing conflict and multiple incidents of shelling and drone attacks, radiation levels at ZNPP have thus far remained within normal limits. The International Atomic Energy Agency (IAEA) has consistently monitored the site, reporting no significant radiation leaks. However, the potential for a catastrophic release of radioactive material continues to loom large, particularly if the plant's safety systems are compromised by further military actions or technical failures.

IAEA Principles: The IAEA has outlined five concrete principles to ensure the plant's safety and security:

- **No attacks from or against the plant**
- **No use of the plant as a military base**
- **No placement of off-site power at risk**
- **Protection of all essential structures from attacks or sabotage**
- **No actions undermining these principles**

The IAEA continues to closely monitor the situation, stressing that the potential danger of a major nuclear accident remains high. Recent drone attacks and military actions in the vicinity of the plant have further exacerbated these risks, highlighting the precarious nature of nuclear safety at ZNPP. The IAEA continues to advocate for the cessation of military activities around the plant and calls for the return of the facility to Ukrainian control to ensure its safety and security.

International Concerns: The international community has expressed grave and growing concerns over the safety and security of ZNPP. The United Nations and various countries have called for the immediate withdrawal of Russian military forces from the plant and the establishment of a demilitarized zone around the facility. The IAEA has maintained a permanent presence at the site to monitor the situation and provide technical support, but restricted access to certain areas complicates efforts to ensure comprehensive safety assessments.

Emergency Response Plans: In response to the heightened risk of a nuclear incident, Ukraine has conducted multiple disaster response drills to prepare for potential emergencies at the Zaporizhzhia plant. These drills involve coordination between various emergency services and aim to enhance the readiness of local authorities and the population to respond effectively to a nuclear disaster. The IAEA has also been involved in supporting Ukraine's emergency preparedness efforts, providing guidance on best practices for nuclear safety and security. However, the ongoing conflict and the presence of military forces at the plant complicate the implementation of these plans, raising concerns about the effectiveness of emergency responses in the event of a real incident.

IAEA Warnings: Rafael Mariano Grossi, Director General of the IAEA, warned the UN Security Council in April 2024 that the prospect of a nuclear accident at ZNPP is "dangerously close." He emphasized that two years of war have significantly compromised nuclear safety at the plant.

Potential Consequences and Comparisons

While the Zaporizhzhia Nuclear Power Plant is not identical to the Chernobyl plant, experts express serious concerns about the potential consequences of a security breach. If the reported explosives were to detonate, it could result in the release of radiation, creating a localized radiation zone with an increased risk of cancer over several decades. However, it is unlikely to result in the same level of destruction witnessed after the Chernobyl disaster in 1986. Unlike Chernobyl, the Zaporizhzhia reactors are in shutdown status and have hard oxide fuel encased in metal, contained within a reinforced concrete structure. Experts suggest that any disaster at Zaporizhzhia would more closely resemble the effects seen at the Three Mile Island incident in Pennsylvania in 1979, where radiation releases had minimal effects on the surrounding populations and environment, rather than the catastrophic scale of Chernobyl or Fukushima.

The safety of ZNPP has become a significant concern due to its continuing occupation by Russian forces amid the ongoing conflict in Ukraine, and the international community remains vigilant. Governments and international organizations have protocols and contingency plans to respond to and mitigate the consequences of any potential incidents. The IAEA's ongoing, close monitoring emphasizes the need for transparency and verification.

So, the Zaporizhzhia Nuclear Power Plant remains a critical concern in the Ukraine conflict. The continued military presence, recent attacks, and challenges in maintaining safety protocols continue to pose significant risks. International efforts to secure the plant and prevent a nuclear incident remain crucial as the conflict persists.

Ukrainian President Zelensky addressed the United Nations General Assembly on September 25th, 2024, emphasizing his concerns over nuclear safety at the Zaporizhzhia Plant and at the other nuclear power plants in his country amid continued Russian bombing.

Conclusion

In this chapter, we have seen that safety is the top priority in the operation of nuclear power plants. Through the implementation of defense in depth, physical barriers, redundant safety systems, advanced technologies, rigorous training, maintenance programs, and emergency response plans, the nuclear industry ensures that these facilities meet the highest safety standards. We can clearly see the nuclear power industry's commitment to protecting the public and the environment.

In a later chapter, I'll be discussing modern nuclear power plant designs that incorporate advanced technologies to enhance safety. For instance, passive safety systems, which rely on natural forces like gravity and convection, are designed to function without the need for human intervention or external power. These systems provide an added layer of safety and increase the plant's resilience to unforeseen events.

We also addressed common concerns about the potential for nuclear power plants to explode like atomic bombs, providing a clear explanation of why such an event is impossible due to fundamental differences in design and operation. We also examined the serious safety concerns at the Zaporizhzhia Nuclear Power Plant amid the ongoing conflict in Ukraine, highlighting the international efforts and challenges in maintaining its security.

With continuous improvements, international cooperation, and a strong safety culture, nuclear power remains one of the safest and most sustainable energy options available today, allowing us to harness the immense potential of nuclear power to combat global warming.

References

1. *5 Facts to know about Three Mile Island.* (2022, May 4). Energy. gov. https://www.energy.gov/ne/articles/5-facts-know-about-three-mile-island

2. Ritchie, H., & Roser, M. (2024, March 18). *What was the death toll from Chernobyl and Fukushima?* Our World in Data. https://ourworldindata.org/what-was-the-death-toll-from-chernobyl-and-fukushima#article-citation

3. *home.* (2024.). NRC Web. https://www.nrc.gov/

4. *Nuclear safety, security and safeguards in Ukraine.* (n.d.). IAEA. https://www.iaea.org/topics/response/nuclear-safety-security-and-safeguards-in-ukraine

5. *Ukraine: Current status of nuclear power installations.* (2024, June 28). Nuclear Energy Agency (NEA). https://www.oecd-nea.org/jcms/pl_66130/ukraine-current-status-of-nuclear-power-installations

Chapter 7

The Environmental Impact of Nuclear Power: Waste Management and Decommissioning

Waste management is central to the nuclear fuel cycle. Although nuclear power generation produces a relatively small volume of waste compared to other energy production forms, the safe and effective management and disposal of radioactive waste are absolutely essential. Advanced technologies, such as deep geological repositories and reprocessing, are being developed to enhance the safety and security of nuclear waste storage and disposal.

The safe disposal of high-level waste (HLW) remains one of the nuclear energy sector's most critical challenges. Despite significant advancements in technology and regulatory frameworks, no country has yet established a fully operational, long-term solution for HLW disposal, including spent fuel. This chapter explores the global status of HLW disposal, focusing on key developments, technological advancements, and ongoing challenges.

We will also examine the environmental impact of nuclear power, with a particular focus on waste management and disposal. Addressing public concerns, including the pervasive "Not In My Backyard" (NIMBY) perspective, is crucial to advancing nuclear energy solutions (See Chapter 13). By understanding the strategies and advancements in waste management, we can better appreciate the nuclear industry's commitment to minimizing environmental impact and ensuring long-term safety. Additionally, we will discuss the decommissioning of reactors at the end of their useful life.

Safety is paramount in the nuclear industry, especially in handling uranium fuel and radioactive waste. Strict safety protocols are mandatory to prevent accidents and protect both people and the environment. Regulatory bodies such as the International Atomic Energy Agency (IAEA) and the U.S. Nuclear Regulatory Commission (NRC) enforce comprehensive safety standards to safeguard nuclear facilities and minimize the risks associated with nuclear fuel and waste.

Types and Management of Radioactive Waste

Classification: Radioactive waste from nuclear power plants is categorized into three main types: low-level waste (LLW), intermediate-level waste (ILW), and high-level waste (HLW). LLW and ILW primarily consist of materials with low levels of radioactivity, such as protective clothing and tools. These types of waste require less stringent safety measures and have shorter time frames for their radioactivity to reach safe levels compared to HLW, which includes highly radioactive materials like spent nuclear fuel.

Waste Management Strategies: The nuclear industry employs a multi-layered approach to manage and dispose of radioactive waste. High-level nuclear waste primarily consists of spent nuclear fuel and waste generated from the reprocessing of spent fuel. This waste remains highly radioactive and thermally hot, necessitating secure and long-term disposal solutions to protect human health and the environment for thousands of years.

Half-Lives and Radioactive Decay: The half-lives of radioactive isotopes in high-level radioactive waste vary depending on the specific elements and isotopes involved. The half-life refers to the period it takes for half of the radioactive material to decay into a stable form. For example, plutonium-239, a commonly found isotope in HLW, has a half-life of approximately 24,000 years. After 24,000 years, half of the original material remains radioactive. This process continues until the levels of radioactivity eventually decay to a point considered safe. Different radioactive isotopes have different half-lives, ranging from extremely short times to exceedingly long ones. Plutonium-239 is of particular concern due to its very long half-life.

Radioactive Decay in Spent Nuclear Fuel

The radioactivity in spent nuclear fuel decays relatively quickly at first, then more slowly over longer time periods.

Here are some key points about the decay of radioactivity in spent fuel:

- In the first few days and weeks after removal from the reactor, the radioactivity decreases rapidly:
- After 1 day, the decay heat (which correlates with radioactivity) falls to about 0.4% of the reactor's operating power.
- After 1 week, it decreases further to about 0.2%.

The rate of decay slows but continues to decrease significantly over years and decades:

- After 40 years, the radioactivity has decreased to about one-thousandth of the level when it was first removed from the reactor.
- After 50 years of storage, the radioactivity has decayed enough that it becomes much easier to handle the waste safely.

Over longer time scales, the decay continues more gradually:

- It takes about 1,000-10,000 years for the radioactivity to decay to the level of the original uranium ore that was mined to produce the fuel.
- The period of greatest radiotoxicity concern extends over about 10,000 years.

The exact decay rate can vary depending on factors like:

- Initial fuel composition
- Reactor burn-up
- Presence of long-lived isotopes like plutonium and americium

It's important to note that while the overall radioactivity decreases significantly, some long-lived isotopes remain radioactive for much longer periods, necessitating careful long-term management and disposal strategies.

Storage: Storage is an integral part of waste management, allowing for the safe, temporary containment of radioactive waste until a permanent disposal solution is implemented. LLW and ILW are stored in specially designed facilities, either on-site or at centralized storage locations. These facilities implement strict safety measures to prevent any release of radiation into the environment.

Technological Advancements: Recent technological advancements have significantly enhanced the handling and storage of HLW. Innovations such as advanced robotics for precise handling, vitrification techniques to immobilize waste in glass or ceramic, and the development of highly durable containers have improved the safety and efficiency of nuclear waste management.

5 Fast Facts on
Spent Nuclear Fuel

1. Spent fuel is a solid and is typically made up of **ceramic pellets in metal rods.**

Spent fuel assemblies inside a dry storage cask. >>>

2. The U.S. has produced roughly **90,000 metric tons** of spent fuel. This could all fit on a football field at a **depth of less than 10 yards** if it could be stacked together.

3. Spent fuel from power reactors is safely and securely stored at more than **70 sites in 35 states.**

Underwater storage at Indian Point in Buchanan, NY.

4. Spent fuel is safely transported across the U.S. with more than **2,500 cask shipments over the last 55 years.**

5. Spent fuel can be recycled. **More than 90% of its potential energy still remains in the fuel.**

Dry storage casks at Dresden Generating Station. >>>

U.S. DEPARTMENT OF **ENERGY** | Office of **NUCLEAR ENERGY** energy.gov/ne

Deep Geological Repositories for High-Level Radioactive Waste

Deep Geological Repositories (DGRs) are regarded as a robust and reliable solution for the long-term disposal of high-level radioactive waste. Locations in stable geological formations, such as deep clay, granite, and salt formations, are considered for hosting these repositories. With isolation time frames of tens of thousands of years, these repositories ensure the safe containment of waste, accounting for the half-lives of the radioactive isotopes present. Through careful planning, scientific analysis, and adherence to stringent safety standards, deep geological repositories offer a responsible and sustainable approach to radioactive waste management.

High-level radioactive waste is typically packaged for very long-term storage using a multi-barrier approach. This approach involves several layers of containment to ensure the safe isolation of the waste. Here is a general overview of the packaging process:

- **Vitrification:** Liquid HLW is converted into a stable glass form, immobilizing the waste and preventing the release of radioactive materials.

- **Direct Packaging:** The waste is placed in a primary container, such as steel or concrete canisters or specially designed casks. These containers are made of materials that can withstand the high levels of radiation emitted by the waste.

- **Shielding:** To further protect against radiation, the containers are often surrounded by additional shielding materials, such as lead or concrete.

- **Secondary Packaging:** The primary containers are then placed in secondary packaging, such as concrete or steel overpacks. This provides an additional layer of containment and shielding.

- **Transportation Containers:** If the waste needs to be transported to a long-term storage facility, it is placed in specially designed transportation containers. These containers are built to withstand accidents and ensure the safe transport of highly radioactive waste.

- **Storage Facility:** The packaged waste is then stored in a dedicated storage facility designed for long-term storage. These facilities are typically located deep underground, in geologically stable formations to provide additional protection against natural disasters and human interference.

The choice of location for deep geological repositories is based on several factors. It involves extensive scientific research and analysis to identify suitable geological formations that can provide the necessary isolation and containment for the waste. These formations should have characteristics such as low permeability, chemical stability, and minimal seismic activity. Common geological formations considered for repository sites include deep clay, granite, and salt formations.

- **Deep Clay Formations:** Deep clay formations, such as those found in France and Belgium, are often considered for repository sites. Clay has low permeability, which limits the movement of water and the migration of radioactive materials. This provides an additional layer of protection against the release of radioactive substances into the environment. The expected time frame for safe disposal in deep clay repositories spans tens of thousands of years.

- **Granite Formations:** Granite formations, like those utilized in the Swedish repository project in Forsmark, offer favorable conditions for deep geological repositories. Granite is a hard and impermeable rock that provides natural barriers against the movement of water and the migration of radioactive materials. The expected time frame for safe disposal in granite repositories is also on the order of tens of thousands of years.

- **Salt Formations:** Salt formations, such as the Waste Isolation Pilot Plant (WIPP) in the United States, are another option for deep geological repositories. Salt has self-sealing properties due to its plastic behavior, which helps to further isolate the waste. The expected time frame for safe disposal in salt repositories can extend into the hundreds of thousands of years.

Expected Time Frames for Safe Disposal

The expected time frames for safe disposal in deep geological repositories must consider the half-lives of the radioactive isotopes in the waste. It is essential to ensure that the radioactivity of the waste decreases to levels deemed safe for human health and the environment over extended periods.

While estimates vary based on repository design and specific waste characteristics, it is generally accepted that several tens of thousands of years are required for radioactive decay to reduce the radioactivity of the waste to a level that poses minimal risk. However, repository designs and safety assessments consider much longer time frames to account for uncertainties and provide a significant margin of safety.

Deep geological repositories are poised to offer a robust and reliable solution for the long-term disposal of high-level radioactive waste. These repositories, located in stable geological formations such as deep clay, granite, and salt, are engineered to ensure safe isolation and containment of waste, factoring in the half-lives of the radioactive isotopes present. Through meticulous planning, scientific analysis, and adherence to stringent safety standards, these repositories represent a responsible and sustainable approach to radioactive waste management.

Two Nordic countries are leading the charge in the development and implementation of deep geological repositories for nuclear waste, while the U.S. has a pilot plant for defense-generated waste:

- **Finland:** The Onkalo repository in Finland is set to become the world's first operational DGR, with waste deposition expected to begin by 2025. This will mark Onkalo as the first deep geological repository licensed for the disposal of used fuel from civil reactors.

- **Sweden:** In Sweden, the Forsmark site has received a construction license, with operations projected to start between 2030 and 2032.

Radioactive Waste Disposal in the United States

The United States has grappled with the challenge of establishing a deep geological repository (DGR) for nuclear power waste for several decades. Despite significant efforts and investments, the country has yet to develop a fully operational DGR for the disposal of high-level radioactive waste and spent nuclear fuel from power reactors.

The Waste Isolation Pilot Plant (WIPP): WIPP is the United States' only deep geological repository for long-lived radioactive waste. Located near Carlsbad, New Mexico, WIPP is situated 2,150 feet underground in a geologically stable salt formation, making it an ideal location for isolating radioactive waste due to the salt's impermeable and self-sealing properties.

WIPP is specifically designed to store defense-generated transuranic waste, including materials contaminated with radioactive elements such as plutonium. Authorized by Congress in 1979, the facility was constructed during the 1980s and began operations in 1999, receiving its first waste shipment from Los Alamos National Laboratory.

The facility disposes of contaminated materials, including clothing, tools, and debris, from various Department of Energy (DOE) sites across the country. WIPP operates under strict environmental and safety regulations to ensure the protection of human health and the environment.

WIPP represents a significant advancement in the long-term management of radioactive waste, demonstrating the feasibility and safety of deep geological disposal for specific types of radioactive materials in the United States.

Yucca Mountain: The suitability of the Yucca Mountain site in Nevada for storing high-level radioactive waste has been a source of intense controversy, driven by a complex web of scientific, technical, and political challenges.

- **Geological and Environmental Issues:** Yucca Mountain is situated in a region that is both seismically and volcanically active. The site's geology, characterized by porous volcanic tuff, raises significant concerns about its ability to safely contain nuclear waste over the very long term. The aquifer beneath Yucca

Mountain drains into the Amargosa Valley, an important agricultural area, which heightens the risk of contamination. Additionally, the site's proximity to Las Vegas and Nellis Air Force Base amplifies worries about potential environmental and safety impacts. For these reasons Yucca Mountain has turned out to be a poor choice for very long term storage of high level radioactive waste.

- **Capacity and Space Limitations:** The statutory design capacity of Yucca Mountain is capped at 77,000 metric tons of heavy metal (MTHM), a figure that falls short of accommodating all of the nation's nuclear waste. With over 70,000 metric tons already stored at various reactor sites and an annual increase of more than 2,000 tons, Yucca Mountain, even it approved soon, would reach capacity by around 2036, leaving the nation without a long-term solution for a substantial amount of radioactive waste.

- **Transportation Risks:** Transporting nuclear waste to Yucca Mountain introduces considerable risks. Proposed routes traverse densely populated areas, affecting over 123 million people across 703 counties in 44 states. This raises serious concerns about the potential for accidents or attacks during transportation, which could result in catastrophic consequences.

- **National Security Concerns:** Concentrating high-level nuclear waste at Yucca Mountain creates a large and potentially vulnerable target for terrorist attacks. The Department of Energy (DOE) plans to store waste above ground at the site for at least 100 years, which only increases the risk of it becoming a target.

- **Regulatory and Legal Challenges:** The suitability of Yucca Mountain has been repeatedly challenged in courts. For instance, the U.S. Court of Appeals ruled against the Environmental Protection Agency's (EPA) initial 10,000-year radiation containment standard, deeming it insufficient. As a result, the EPA was forced to revise its standards to address these legal challenges.

- **Political and Public Opposition:** The Yucca Mountain project has faced fierce opposition from Nevada residents, political leaders, and indigenous groups, including the Western Sho-

shone. This opposition is rooted in the belief that it is unjust to store nuclear waste in a state without nuclear power plants and deep concerns about the site's safety and environmental impact. Political dynamics, including the influence of prominent Nevada politicians, have also played a significant role in halting the project's progress.

In summary, the suitability of Yucca Mountain as a nuclear waste repository remains deeply contested. These unresolved issues have led to ongoing debates and legal battles over its future, leaving the nation at a crossroads in its quest for a long-term nuclear waste solution. Another, more suitable site (or sites) must be found!

Reprocessing Spent Nuclear Fuel

Reprocessing spent nuclear fuel is a critical process that not only reduces the volume of high-level radioactive waste requiring long-term disposal but also recovers valuable materials like uranium and plutonium for reuse in nuclear reactors. During reprocessing, spent nuclear fuel undergoes chemical treatment to separate its components, typically involving dissolution in acid and the extraction of valuable elements. Uranium and plutonium are the primary targets due to their potential for reuse, particularly in the production of mixed oxide (MOX) fuel—a blend that can power certain types of reactors.

The advantages of reprocessing are clear: It allows for the recycling of valuable materials, thus reducing reliance on newly mined uranium and plutonium. Moreover, it enhances overall fuel efficiency and energy output from nuclear power plants, contributing to a more sustainable energy system.

However, reprocessing also presents significant challenges and concerns. Managing separated plutonium is particularly problematic, as plutonium-239, a key isotope, can be used in nuclear weapons. This raises serious proliferation concerns, requiring stringent safeguards and security measures to prevent unauthorized access.

Additionally, reprocessing facilities pose their own environmental and safety risks. The chemical processes involved generate substantial quantities of highly radioactive liquid and solid waste, which demand robust containment, proper disposal, and careful long-term management to protect workers, the public, and the environment.

Due to these complexities, countries have adopted differing approaches. France and Russia, for example, have established commercial reprocessing facilities, with France's Orano Facility having reprocessed over 40 metric tons of used nuclear fuel, demonstrating the feasibility of large-scale operations. In contrast, the United States has limited reprocessing activities, though the Biden administration is now reassessing these technologies with a renewed sense of urgency, particularly in the context of advanced reactor designs.

The decision to pursue reprocessing versus relying solely on direct disposal of spent nuclear fuel is multifaceted, requiring a careful balance of technical, environmental, economic, and non-proliferation factors.

Promising technological advancements are underway to reduce the radiotoxicity and volume of high-level waste. These involve converting long-lived radioisotopes into shorter-lived or stable isotopes through methods like reactor or accelerator transmutation, which aim to further minimize human exposure and environmental contamination.

Radioactivity in Coal-fired Power Plant Residue

While nuclear power often draws the spotlight when discussing radioactive waste, coal-fired power plants also produce significant radioactive residues. In the United States, the Environmental Protection Agency (EPA) enforces stringent regulations to manage coal combustion residuals—commonly referred to as coal ash—safeguarding communities, groundwater, and the environment from contamination.

Most people are unaware that coal ash contains radioactivity. This radioactivity originates from naturally occurring radioactive materials in coal, such as uranium, thorium, and their decay products, including radium and radon. When coal is burned, these materials become concentrated in byproducts like fly ash, bottom ash, and flue gas desulfurization sludge, often resulting in higher radioactivity levels than in the original coal. Although these levels are much lower than those in nuclear waste, they can still pose significant environmental and health risks.

In comparison, nuclear waste is exponentially more radioactive—by factors of 100,000 to 1,000,000—than coal ash, highlighting the advanced containment and management strategies required for its safe disposal. Yet, if improperly managed, coal ash can leach radioactive materials into water sources, leading to contamination risks. Additionally, radon gas released during coal combustion is a public health concern due to its link to increased lung cancer risk.

Coal-fired power plants also require substantial land for waste storage. Fly ash and bottom ash are typically stored in ash ponds or landfills, some of which span several hundred acres. Strict regulatory compliance is essential to prevent groundwater contamination and ensure proper waste containment. Although some coal ash is recycled—such as fly ash used in construction materials—other types, like bottom ash, often require long-term storage in landfills.

Advancements in pollution control technologies and the shift to cleaner energy sources have reduced the environmental impact of coal-fired power plant residues. Nonetheless, the management of coal ash and its radioactivity remains crucial to protecting both environmental and human health.

What Happens to Old Nuclear Reactors? Life Extension and Decommissioning

Before we move on from the topic of nuclear power's effects on the environment, we must take this crucial matter into consideration. Nuclear power reactors can have their operational lifetimes extended well beyond their original design through a process known as "life extension" or "long-term operation" (LTO). This approach has gained momentum as many reactors worldwide approach or surpass their initially licensed operating periods. Here's how older nuclear reactors can be upgraded to continue delivering reliable power:

License Renewal: In the United States, nuclear plants are initially licensed for 40 years. The Nuclear Regulatory Commission (NRC) allows these plants to apply for license renewals in 20-year increments.

Some plants have already received approval to operate for up to 80 years, showcasing the viability of extended operation.

Equipment Replacement and Upgrades: Key components that undergo wear and tear over time are replaced or refurbished to maintain safe and efficient operations. These upgrades typically include:

- Steam generators
- Reactor vessel heads
- Turbine rotors and blades
- Control rod drive mechanisms
- Electrical systems and instrumentation

Safety System Enhancements: Upgrades to safety systems are critical to meeting evolving regulatory standards and integrating lessons learned from past incidents, such as the Fukushima accident. These enhancements often involve:

- Improved emergency power systems
- Enhanced flood and seismic protection
- Upgraded fire protection systems

Digital Instrumentation and Control: Many older plants are transitioning from analog to digital instrumentation and control systems, which significantly improves reliability and operational efficiency.

Power Uprates: Some reactors undergo power uprates to increase their electrical output by enhancing efficiency or slightly boosting reactor power levels. This is achieved through better fuel designs, more efficient turbines, or upgraded instrumentation that allows the plant to operate closer to its design limits.

Structural Integrity Assessments: Critical structures, such as the reactor pressure vessel and containment, undergo detailed evaluations to ensure they can withstand the stresses of extended operation. Advanced non-destructive testing techniques are often employed in these assessments.

Aging Management Programs: Comprehensive aging management programs are implemented to monitor and address the effects of aging on plant systems, structures, and components. This includes regular inspections, testing, and preventive maintenance.

Fuel Design Improvements: Advancements in fuel designs not only improve efficiency but also allow for longer fuel cycles and enhance safety margins.

Knowledge Management: As the original workforce retires, it's essential to capture and transfer their knowledge to the next generation of workers. Effective programs are in place to ensure this critical information is not lost.

Economic Assessments: Detailed economic analyses are conducted to justify life extension investments, weighing them against alternatives such as building new plants or transitioning to other energy sources.

The decision to extend the life of a nuclear plant is a complex one, requiring careful consideration of safety, economic, and regulatory factors. While many plants have successfully implemented life extension programs, others have faced early retirement due to economic pressures or technical challenges.

It's worth noting that life extension can be a highly cost-effective strategy to maintain low-carbon electricity generation, as the capital costs of the plant have already been paid off. However, significant investment and rigorous safety evaluations are necessary to ensure the continued safe operation of these extended-life reactors.

The Palisades nuclear power plant in Michigan, which was shut down in 2022 due to financial challenges, is set to make history as the first decommissioned reactor in the U.S. to restart. Pending approval from the Nuclear Regulatory Commission (NRC), Palisades is expected to resume operations by late 2025, generating enough energy to power over 800,000 homes. Holtec, the plant's owner, has ambitious plans to expand the site's capacity by introducing small modular reactors, potentially doubling its output and reinforcing the role of nuclear energy in the fight against climate change.

Decommissioning Old Reactors

As the world transitions towards greener and more sustainable energy sources, decommissioning old nuclear reactors is becoming increasingly essential. Decommissioning refers to the safe dismantling of nuclear power plants after they have reached the end of their useful life. This section explores the environmental challenges and critical safety measures involved in this complex and necessary process.

Challenges in Decommissioning

- **High Costs:** Decommissioning is an expensive process, often requiring method and the specific conditions of the site.

- **Long Time Frames:** The decommissioning process can take several decades, especially when using the SAFSTOR method, which allows for radioactive decay before dismantling begins. This extended timeline complicates planning and funding.

- **Waste Management:** Managing and disposing of radioactive waste safely is a significant challenge. High-level waste requires secure, long-term storage solutions, while low and intermediate-level waste must be carefully handled to prevent environmental contamination.

Effective Methods for Decommissioning Old Nuclear Reactors

Decommissioning old nuclear reactors is a multi-faceted process that requires meticulous planning and execution to ensure safety and minimize environmental impact. Several methods have been developed and refined over the years to handle the decommissioning of nuclear facilities effectively. The primary methods include Immediate Dismantling (DECON), Deferred Dismantling (SAFSTOR), and Entombment (ENTOMB). Each method has its advantages and disadvantages, which must be carefully weighed based on the specific circumstances of the reactor being decommissioned.

Decommissioning Methods for Nuclear Reactors

Method	Description	Steps Involved	Advantages	Disadvantages
Immediate Dismantling (DECON)	Involves prompt dismantling and decontamination of the facility shortly after it has been permanently shut down.	1. **Removal of Nuclear Fuel:** Removes spent nuclear fuel and places it into a used fuel pool or dry storage containers. 2. **Dismantling and Decontamination:** removes or decontaminates equipment, structures, and components with radioactive materials. 3. **Waste Management:** Manages radioactive waste by storing waste in deep geological repositories and decontaminating or transporting low and intermediate-level wastes to waste-processing facilities.	- **Immediate Risk Reduction:** Reduces radiation exposure risk by promptly removing radioactive materials. - **Utilization of Existing Workforce:** Uses the existing workforce familiar with the facility.	- **High Initial Costs:** Requires significant upfront investment. - **Intensive Planning and Execution:** Demands meticulous planning and coordination.
Deferred Dismantling (SAFSTOR)	Maintains the facility in a safe and stable condition for an extended period, allowing radioactivity to decay naturally before dismantling begins. *Note: After 50 years the total radiation levels are less than 1% of the original radioactivity.*	1. **Safe Storage:** Maintains and monitors the facility to ensure it remains safe. 2. **Decay Period:** Leaves the facility in a safe state for 30 to 50 years, allows radioactivity levels to decrease naturally. 3. **Final Dismantling:** Dismantles the facility and decontaminates the site after the decay period, reducing radiation hazards and potentially lowering decommissioning costs.	- **Reduced Radiation Hazard:** Allows time for radioactive decay, reducing radiation levels and hazards. - **Potential Cost Savings:** Can result in lower decommissioning costs due to reduced radiation levels and advancements in technology.	- **Extended Timeframe:** Delays final decommissioning and site reuse for several decades. - **Future Uncertainties:** Potentially higher future costs due to inflation, changing regulations, and ongoing maintenance and monitoring.
Entombment (ENTOMB)	Involves encasing the radioactive components of the facility in a long-lasting material, such as concrete, effectively isolating them from the environment.	1. **Partial Entombment:** Dismantles above-ground structures and entombs below-ground structures. 2. **Complete Entombment:** Encapsulates the entire facility and monitors it indefinitely to ensure containment remains intact.	- **Lower Initial Costs:** Can be less expensive than complete dismantling - **Safer Execution:** Reduces risk to workers by minimizing handling of highly radioactive materials.	- **Indefinite Monitoring:** Requires ongoing monitoring and maintenance. - **Permanent Site Use Restrictions:** The site remains unusable for other purposes and must be managed as a long-term radioactive waste site.

Decommissioning nuclear reactors requires the utmost care to minimize environmental impact. The primary concerns are the management of radioactive waste, the prevention of contamination, and the restoration of the site for future use.

Environmental Aspects of Decommissioning Nuclear Reactors

Radioactive Waste Management: One of the most significant environmental challenges in decommissioning is the safe handling of radioactive waste. High-level waste, such as spent nuclear fuel, must be securely stored and isolated from the environment for thousands of years. Intermediate- and low-level radioactive waste, while less hazardous, still demands meticulous handling. These materials are often stored in specially designed facilities where they can safely decay over time.

The goal is clear: to ensure that no radioactive material escapes into the environment, thereby protecting human health and ecosystems.

Contamination Prevention: Preventing contamination during decommissioning is crucial. This involves the decontamination of equipment, structures, and contaminated soil. Techniques such as chemical cleaning, abrasive blasting, and high-pressure water jets are employed to remove radioactive materials from surfaces. Additionally, contaminated soil may be excavated and treated or disposed of in a controlled manner.

Monitoring and controlling the spread of contamination is vital. This includes regular environmental assessments and the use of barriers to prevent the migration of radioactive particles. By maintaining strict control over contamination, the environmental impact of decommissioning can be significantly reduced.

Site Restoration: Once decommissioning is complete, the site must be restored to a condition that allows for safe future use. This process, known as site remediation, involves the complete removal of all radioactive materials and the restoration of the land to its natural state or for other productive uses. The objective is to ensure that the site poses no risk to human health or the environment. Site restoration may also include re-establishing natural habitats, planting vegetation, and monitoring the site to ensure that no residual contamination remains. This not only mitigates the environmental impact of the decommissioning process but also contributes to broader goals of environmental conservation and sustainability.

Safety Measures in Decommissioning

As with all aspects of handling radioactive materials, safety is paramount in the decommissioning of nuclear reactors. The process involves several critical safety measures to protect workers, the public, and the environment.

Regulatory Framework: Decommissioning is governed by a stringent regulatory framework established by national and international bodies, such as the International Atomic Energy Agency (IAEA). These regulations set out the standards and procedures that must be followed to ensure safety throughout the decommissioning process. Compliance with these regulations is mandatory and subject to rigorous oversight.

Worker Safety: Protecting the health and safety of workers involved in decommissioning is a top priority. Workers are provided with personal protective equipment, including radiation suits, gloves, and respirators, to shield them from exposure. Additionally, strict protocols are in place for monitoring radiation levels and limiting the time workers spend in high-radiation areas. Each worker is individually monitored to ensure that any radiation doses received are kept within acceptable safety standards, further ensuring their protection throughout the decommissioning process.

Training and education are also crucial components of worker safety. Workers receive comprehensive training on radiation protection, emergency procedures, and the safe handling of radioactive materials. This ensures that they are well-prepared to conduct their tasks safely and effectively.

Emergency Preparedness

Despite all precautions, the potential for accidents or unexpected events cannot be entirely eliminated. Therefore, robust emergency preparedness plans are essential. These plans outline the actions to be taken in the event of a radiation release or other emergencies. They include evacuation procedures, communication protocols, and coordination with local authorities and emergency services.

Regular drills and exercises are conducted to ensure that all personnel are familiar with emergency procedures and can respond quickly and effectively in the event of an incident. This preparedness helps minimize the impact of any potential accidents and protect both workers and the public.

NRC's Role in Safeguarding Public Health and the Environment in the United States

In the United States, the Nuclear Regulatory Commission (NRC) ensures the safe decommissioning of nuclear power reactors with a focus on public health, safety, and environmental protection. Governed by strict regulations, the NRC's oversight begins at plant shutdown and continues through to the final site release. This framework ensures compliance at every stage.

The NRC allows nuclear plant operators to choose between two primary decommissioning strategies:

- **DECON**: Immediate dismantling of the facility
- **SAFSTOR**: Deferred dismantling, where the plant is maintained and monitored in a safe condition for several decades before eventual cleanup.

These strategies allow flexibility, but both are subject to rigorous oversight to ensure adherence to safety and environmental standards.

One of the NRC's critical responsibilities is to guarantee that nuclear plant operators maintain adequate financial resources to cover the entire cost of decommissioning. Operators must report on their decommissioning funds every two years, and the NRC monitors these funds closely to prevent financial shortfalls that could compromise safety or delay cleanup efforts.

At the core of the NRC's decommissioning oversight is ensuring that residual radioactivity at a decommissioned site is reduced to levels that meet regulatory standards. After cleanup, the NRC reviews the Final Status Survey Report submitted by the operator, which documents the site's radiological conditions. If the site meets NRC requirements, the facility's license is terminated, and the site can be released for unrestricted or restricted use depending on the level of contamination.

Throughout the decommissioning process, the NRC actively engages with the public. Public meetings, comment periods, and hearings on critical decisions, such as license amendments, offer opportunities for transparency and community involvement. This ensures that local communities are informed and involved in decisions that may impact their environment and health.

So, we see that through detailed oversight, financial monitoring, and public engagement, the NRC safeguards both current and future generations from potential risks associated with decommissioned nuclear sites.

History of Decommissioning Nuclear Power Reactors

Decommissioning nuclear reactors is a critical process that involves safely dismantling nuclear facilities that have reached the end of their operational lives. This process is essential for ensuring environmental safety and public health. The history of decommissioning power reactors varies across the United States and around the world, reflecting different regulatory environments, technological capabilities, and historical contexts.

The United States has a significant history of nuclear reactor decommissioning, driven by an aging fleet of reactors and evolving regulatory standards. As of 2017, ten commercial nuclear reactors in the U.S. had been successfully decommissioned, with another twenty in various stages of the process.

Notable Decommissioned Reactors in the U.S.

1. **Haddam Neck Plant (Connecticut Yankee):** This 619 MW reactor in Connecticut was shut down in 1997 and decommissioned using the DECON method, with the process completed in 2007 at a cost of $893 million.

2. **Trojan Nuclear Plant:** Located in Oregon, this 1180 MW reactor was shut down in 1993. The dismantling process included the removal of steam generators and the reactor vessel, with the site released for unrestricted use in 2005, excluding the spent fuel storage area.

3. **Maine Yankee:** This 860 MW reactor in Maine was shut down in 1996. Decommissioning was completed in 2005, with the site largely returned to unrestricted use.

4. **Big Rock Point:** This 72 MW reactor in Michigan was shut down in 1997, and the site was largely returned to greenfield status by 2007, with some land still used for dry cask storage of spent fuel.

Globally, decommissioning efforts have varied based on national policies, reactor types, and historical incidents. Approximately 200 commercial, experimental, or prototype reactors, over 500 research reactors, and several fuel cycle facilities have been retired from operation worldwide.

Other Notable Decommissioning Projects

1. **Savannah River Site (USA):** The P and R reactors at this site in South Carolina were decommissioned using a partial entombment method, involving the grouting of below-ground structures. This approach significantly reduced decommissioning costs compared to traditional methods, making it a more efficient option for specific reactor types.

2. **Three Mile Island Unit 2 (USA):** After the partial meltdown in 1979, Unit 2 was placed in SAFSTOR. The reactor remains in a monitored storage state, awaiting the decommissioning of Unit 1 to facilitate a combined dismantling process. (Recall that the undamaged Unit 1 continued to operate until 2019.)

3. **Hallam Nuclear Reactor (USA):** This reactor in Nebraska was shut down in 1964 and entombed by encasing the reactor in concrete. While less common, this method has proven effective for certain types of reactors, providing a long-term containment solution.

4. **Chernobyl Reactor 4 (Ukraine):** Following the catastrophic accident in 1986, Reactor 4 was entombed in a concrete sarcophagus to contain the radiation. This ENTOMB method was chosen due to the extreme contamination levels and the urgent need for immediate containment.

5. **Fukushima Daiichi Site (Japan):** In the wake of the 2011 nuclear accident, substantial efforts have been made to stabilize the damaged reactors. Units 1-3 experienced partial meltdowns due to cooling system failures, necessitating continuous cooling and containment to prevent further radioactive releases. Plans are in place to remove all fuel rods by 2031 and extract molten fuel debris by 2050, with full decontamination and decommissioning expected to take 30 to 40 years. The reactors are currently in "cold shutdown," maintaining stable temperatures but remaining highly radioactive and inaccessible to the public. This site will require long-term management and ongoing remediation efforts.

6. **San Onofre (USA):** The decommissioning of the San Onofre Nuclear Generating Station in California began in 2020, seven years after the plant ceased operations in 2013. The dismantlement process is projected to take 8 to 10 years, with the removal of its iconic domes scheduled for 2027-2028. However, 123 canisters of spent nuclear fuel remain on-site, awaiting federal removal before the site can be fully cleared.

7. **Diablo Canyon (USA):** The Diablo Canyon Nuclear Plant, initially set for decommissioning in the mid-2020s, received a temporary reprieve in 2022 when California lawmakers passed a bill allowing the plant to continue operating until 2029 or 2030. Both of its nuclear reactors are still in operation, providing power to the state's energy grid while Pacific Gas & Electric prepares for decommissioning and simultaneously upgrades the plant's infrastructure. The Nuclear Regulatory Commission is also reviewing a 20-year license extension application for the plant. Despite these developments, California law continues to prohibit the construction of new nuclear power plants until a permanent national solution for spent fuel storage is established.

In summary, the decommissioning process involves several critical steps, including the removal of nuclear fuel, decontamination of equipment and structures, and the disposal of radioactive waste. The choice of decommissioning strategy—DECON, SAFSTOR, or EN-

TOMB—depends on various factors, such as contamination levels, available funding, and future site use plans.

- Immediate Dismantling (DECON) offers the advantage of prompt risk reduction and site reuse but comes with high initial costs.

- Deferred Dismantling (SAFSTOR) allows for significant natural radioactive decay, potentially lowering costs and hazards, but delays site reuse.

- Entombment provides a cost-effective and safer alternative but requires indefinite site monitoring and restricts future site use.

Each method must be carefully evaluated to ensure the safe, efficient, and environmentally responsible decommissioning of nuclear reactors.

The history of decommissioning nuclear reactors in the U.S. and worldwide highlights the complexity and critical importance of this process. Effective decommissioning ensures that former nuclear sites are safely managed and can be repurposed for other uses, contributing to environmental protection and public health. As more reactors reach the end of their operational lives, the lessons learned from past decommissioning projects will be invaluable in guiding future efforts.

Conclusion

Nuclear power generation, while producing radioactive waste, offers significant environmental advantages when managed responsibly. The volume of nuclear waste is small compared to other industrial toxic wastes, and safe disposal methods, particularly in deep geological repositories, have been extensively studied and proven technically feasible. With proper containment and isolation from the environment over very long timescales, radiological risks can be minimized to negligible levels.

Given nuclear power's low greenhouse gas emissions, it is an essential component of the clean energy transition needed to combat global warming. By continuing to improve waste management techniques and implementing permanent geological repositories for high-level waste, the nuclear industry can ensure that waste issues do not hinder its crucial role in providing reliable, carbon-free baseload electricity to complement renewable sources. With sound waste disposal solutions in place, nuclear power's environmental advantages become even more compelling as the world seeks sustainable pathways to decarbonize the energy sector.

References

1. *Radioactive waste.* (n.d.). NRC Web. https://www.nrc.gov/waste.html

2. Sabine Hossenfelder. (2022, November 26). *Nuclear waste is not the problem you've been made to believe it is* [Video]. YouTube. https://www.youtube.com/watch?v=aDUvCLAp0uU

3. *Storage and disposal of radioactive waste - World Nuclear Association.* (n.d.). https://world-nuclear.org/information-library/nuclear-fuel-cycle/nuclear-waste/storage-and-disposal-of-radioactive-waste

4. *Deep geological repository development to be strengthened with Horonobe International Project.* (2023, April 25). Nuclear Energy Agency (NEA). https://www.oecd-nea.org/jcms/pl_80835/deep-geological-repository-development-to-be-strengthened-with-horonobe-international-project

5. *NEA Publications: ISSUE BRIEF No. 3 - THE DISPOSAL OF HIGH-LEVEL RADIOACTIVE WASTE.* (1989, January). NEA. https://www.oecd-nea.org/brief/brief-03.html

6. Yu, A. (2023, June 26). Where can the U.S. put 88,000 tons of nuclear waste? *WHYY.* https://whyy.org/segments/us-nuclear-waste-store-tons/

7. Vattenfall. (2023, August 29). *Finland to open the world's first final repository for spent nuclear fuel.* Vattenfall. https://group.vattenfall.com/press-and-media/newsroom/2023/finland-to-open-the-worlds-first-final-repository-for-spent-nuclear-fuel

8. Portuondo, N. (2024, April 11). *The return of Yucca Mountain? GOP floats waste site's revival.* E&E News by POLITICO. https://www.eenews.net/articles/the-return-of-yucca-mountain-gop-floats-waste-sites-revival/

9. INFOGRAPHIC: 5 Fast Facts about Spent Nuclear Fuel | Department of Energy

10. Kramer, D. (2024, February 1). US takes another look at recycling nuclear fuel. *Physics Today, 77*(2), 22–25. https://doi.org/10.1063/pt.mrdf.volt

11. *Radioactive wastes from coal-fired power plants | US EPA.* (2024, July 9). US EPA. https://www.epa.gov/radtown/radioactive-wastes-coal-fired-power-plants

12. *First nuclear reactor restart in US history secures $1.5B in federal funding.* (n.d.). https://www.msn.com/en-us/money/companies/first-nuclear-reactor-restart-in-us-history-secures-1-5b-in-federal-funding/vi-AA1rykS-f?ocid=msedgntp&pc=HCTS&cvid=d4a2727fb99b4f-f7aefc8669376c52a8&ei=15#details

13. *Decommissioning nuclear facilities - World Nuclear Association.* (2022, May 3). https://world-nuclear.org/information-library/nuclear-fuel-cycle/nuclear-waste/decommissioning-nuclear-facilities

14. NRCgov. (2022, June 23). *NRC Decommissioning nuclear power plants - Updated 2022* [Video]. YouTube. https://www.youtube.com/watch?v=picQ5e-m_bA

15. INTERNATIONAL ATOMIC ENERGY AGENCY. (2023). *Global status of decommissioning of nuclear installations.* IAEA. https://www.iaea.org/publications/15197/global-status-of-decommissioning-of-nuclear-installations

16. INTERNATIONAL ATOMIC ENERGY AGENCY. (2021). *Occupational radiation protection during the decommissioning of nuclear installations.* IAEA. https://www.iaea.org/publications/14858/occupational-radiation-protection-during-the-decommissioning-of-nuclear-installations

Chapter 8

Advanced Nuclear Reactors: Exploring the Newest Generation of Power Sources

As we confront the pressing challenges of global warming, the need for safe, efficient, and sustainable energy solutions has never been more urgent. Generation IV nuclear reactors represent the cutting edge of nuclear technology, designed to overcome the limitations of earlier reactor designs, and meet the growing global demand for clean energy.

Generation IV Reactors: The Future of Nuclear Power

These advanced reactors are not just theoretical; they are being actively developed under the auspices of the Generation IV International Forum, which identified six promising reactor technologies:

- **Molten Salt Reactor (MSR)**
- **Sodium-cooled Fast Reactor (SFR)**
- **Gas-cooled Fast Reactor (GFR)**
- **Lead-cooled Fast Reactor (LFR)**

- **Supercritical Water-cooled Reactor (SCWR)**

- **Very High Temperature Reactor (VHTR)**

Generation IV reactors incorporate advanced safety features, notably passive safety systems that eliminate the need for active controls or human intervention to prevent accidents. These systems harness natural physical principles—such as gravity, natural convection, and high-temperature resistance—creating an operational environment that is inherently safer.

Passive Safety Systems

Passive safety systems in these advanced reactors capitalize on natural physical phenomena and the inherent design features of the reactors to ensure safety. These systems function without requiring external power sources or active intervention, offering a robust, fail-safe mechanism that enhances overall reactor safety.

Here's how these systems function:

- **Natural Circulation Cooling:** Many advanced reactor designs utilize natural convection to circulate coolant through the reactor core, removing heat even if pumps fail. This process relies on the density differences in the coolant as it heats up and cools down.

- **Gravity-Driven Safety Systems:** Some designs incorporate elevated water tanks that can flood the reactor core using only gravity in an emergency, providing cooling without the need for pumps.

- **Passive Heat Removal:** Advanced reactors often have systems to passively transfer decay heat from the core to the environment using natural circulation of non-radioactive air or water.

- **Inherent Negative Reactivity Feedback:** The physics of the reactor core is designed so that as temperature increases, reactivity decreases, naturally limiting power excursions and enhancing safety.

- **Accident-Tolerant Fuels:** New fuel designs can better withstand high temperatures and retain fission products, enhancing safety margins and reducing the risk of radioactive release.

- **Passive Pressure Relief:** Safety valves that operate based on pressure differentials can relieve excess pressure without the need for powered systems, providing an additional layer of safety.

- **Walk-Away Safe Designs:** Some advanced reactors are designed to safely shut down and cool themselves indefinitely without any operator actions or external power, ensuring long-term safety.

- **Reduced Reliance on Active Components:** By minimizing the need for pumps, valves, and other active components, these designs reduce potential failure points and increase overall reliability.

- **Extended Coping Times:** Passive systems are often designed to maintain safe conditions for much longer periods (days or weeks) compared to active systems in current reactors, providing ample time to address any issues.

- **Simplified Designs:** Many advanced reactors have fewer components overall, reducing complexity and the likelihood of failure.

These passive safety features aim to make advanced reactors more resilient to various accident scenarios, including station blackouts and loss-of-coolant accidents, while reducing reliance on human intervention and backup power systems. By harnessing fundamental physical principles like gravity, natural circulation, and inherent nuclear characteristics, these systems provide multiple layers of defense-in-depth to prevent core damage and radioactive release.

Enhanced Efficiency: These reactors maximize the energy extracted from nuclear fuel, which not only reduces the amount of fuel required but also minimizes nuclear waste. Remarkably, some designs can convert up to 95% of the energy in the fuel into usable electricity, compared to less than 5% in traditional reactors. This improved efficiency extends the time between refueling, reducing the amount of spent fuel generated and the frequency of refueling outages.

Waste Minimization: Some Generation IV reactors are specifically designed to produce less long-lived radioactive waste. Fast reactors, for example, can consume existing nuclear waste as fuel, transforming long-lived isotopes into shorter-lived fission products. This capability significantly reduces the volume and toxicity of nuclear waste.

Proliferation Resistance: By incorporating advanced fuel cycles and reactor designs that do not produce weapons-grade materials, Generation IV reactors make it more challenging to divert nuclear materials for weapon production. This is a critical feature in ensuring global security while expanding nuclear energy.

Economic Competitiveness: Although the initial costs of building Generation IV reactors can be high, their improved fuel efficiency, reduced waste management costs, and potential for co-generation of heat and electricity position them as economically competitive options over their lifecycle.

Key Generation IV Reactor Designs

In 2021, the Office of Nuclear Energy published a report identifying "3 Advanced Reactor Systems to Watch by 2030." They include:

Sodium-cooled Fast Reactors (SFRs): SFRs are advanced nuclear reactors that utilize liquid sodium as a coolant, which allows them to operate at higher temperatures and achieve better thermal efficiency compared to traditional reactors that use water. These reactors are designed to optimize fuel use by converting fertile isotopes into fissile fuel that can sustain a nuclear reaction, effectively creating a closed fuel cycle. Additionally, SFRs have the capability to use nuclear waste as fuel, helping to reduce the amount of long-lived radioactive waste. The reactors incorporate passive safety features, such as systems that automatically prevent overheating without requiring active intervention. They also include robust containment measures to manage the reactive properties of sodium. While SFRs represent a promising and sustainable energy solution, ongoing research is needed to address technical challenges related to materials and safety engineering.

Molten Salt Reactors (MSRs): MSRs use a liquid mixture of salts as both coolant and fuel, offering several significant advantages. Operating at temperatures of 600-700°C, MSRs achieve enhanced thermal efficiency, not only enabling efficient electricity generation but also creating the potential for hydrogen production. These reactors possess inherent safety features, such as low-pressure operation and the ability to drain fuel into passively cooled tanks during emergencies. MSRs also offer fuel cycle flexibility, capable of utilizing various cycles, including thorium-uranium and uranium-plutonium. However, while MSRs hold great promise for future clean energy systems, ongoing research is crucial to overcome the challenges associated with developing materials that can withstand the high-temperature, corrosive salt environment.

High-Temperature Gas-cooled Reactors (HTGRs): HTGRs utilize helium gas as a coolant and graphite as a moderator, allowing them to operate at extremely high temperatures, often exceeding 750°C. These reactors achieve remarkable thermal efficiency, generating more electricity per unit of fuel and minimizing waste. HTGRs are particularly well-suited for hydrogen production and can provide the high temperatures required for various industrial processes, reducing reliance on fossil fuels. Their advanced safety features and the use of inert helium as a coolant position them as a promising solution for a sustainable, low-carbon energy future.

And given their advantages, we should also be paying close attention to small modular reactors:

Small Modular Reactors (SMRs): SMRs are compact, modular nuclear reactors that can be built in factories and transported to sites. They offer several advantages over large nuclear plants, including scalability and reduced initial capital investment. SMRs will be discussed in detail in the next chapter.

Thorium Reactors: A Promising Future Alternative?

Thorium reactors present a compelling alternative to traditional uranium-based nuclear reactors, addressing many of the concerns associated with nuclear energy:

Abundance and Efficiency: Thorium is three times more abundant in the Earth's crust than uranium. Remarkably, a single ton of thorium can produce as much energy as 200 tons of uranium or 3,500,000 tons of coal.

Safety: Thorium reactors, particularly liquid fluoride thorium reactors (LFTRs), operate at atmospheric pressure, which reduces the risk of explosions. These reactors include inherent failsafe measures, such as a fusible plug that melts and drains the reactor in case of overheating, preventing meltdowns.

Reduced Nuclear Waste: Thorium reactors produce significantly less long-lived radioactive waste. Moreover, the waste they do produce remains hazardous for a shorter period—typically a few hundred years compared to tens of thousands of years for uranium waste.

Proliferation Resistance: Thorium reactors produce uranium-233, which is difficult to weaponize due to the presence of uranium-232, a strong gamma emitter that complicates handling and detection.

Challenges: Developing thorium reactors requires significant investment in research, development, and infrastructure. Additionally, thorium's fuel cycle is complex, requiring a neutron source such as enriched uranium or plutonium to sustain the reaction.

Projects and Advanced Reactor Developments

Projects in the United States

- **EBR-II:** Operated from 1964 to 1994 at Argonne National Laboratory, this project demonstrated the feasibility of a closed fuel cycle and passive safety features, which have significantly influenced modern sodium-cooled fast reactor (SFR) designs.

- **Natrium Reactor:** Bill Gates' TerraPower in partnership with GE Hitachi has started construction on the Natrium demonstration project in Wyoming. The Natrium design, a 345 MWe sodium-cooled fast reactor with an integrated molten salt-based energy storage system, aims to enhance grid flexibility by boosting output to 500 MWe when necessary. However, the project has faced delays, primarily due to challenges in securing high-assay low-enriched uranium (HALEU) fuel, a key component for its operation. Despite these setbacks, the project is moving forward, with anticipated completion and operational testing now targeted for the early 2030s.

- **Plant Vogtle:** The Vogtle nuclear plant in Georgia houses four reactors. Units 1 and 2 began operation in 1987 and 1989, respectively. The newly constructed Units 3 and 4, the first new nuclear units in the U.S. in over 30 years, employ passive safety systems that can automatically shut down without operator action or external power if necessary.

- **ARC-100:** The ARC-100, a small modular reactor designed by Advanced Reactor Concepts LLC and based on the EBR-II, represents a promising step forward in nuclear technology. With a capacity of 100 MWe, it's engineered for factory production and on-site assembly, boasting passive safety features and extended refueling intervals. However, despite the technical innovations, the ARC-100 project is encountering significant challenges. Initially slated for deployment by 2030, the timeline has been pushed to 2035 due to delays and changes in leadership. The project, in partnership with New Brunswick Power, is currently under regulatory review by the Canadian Nuclear Safety Commission for a potential site at Point Lepreau, Canada. Furthermore, ARC has teamed up with Korea Hydro & Nuclear Power to explore global deployment opportunities. Although progress has been made in regulatory approval and international collaboration, these obstacles underscore the difficulties inherent in advancing nuclear technology on a global scale.

- **U.S. Government Support and Policies:** The Biden Administration is heavily investing in nuclear energy through initiatives like the Advanced Reactor Demonstration Program and the Inflation Reduction Act, which provides tax credits for zero-greenhouse-gas-emitting electricity generation. The U.S. Department of Energy (DOE) is working to establish a domestic supply of high-assay low-enriched uranium to support advanced reactor technologies.

- **U.S. Regulatory and Licensing Progress:** The U.S. Nuclear Regulatory Commission (NRC) is streamlining its licensing processes for advanced reactors, developing new frameworks and rulemaking to support the deployment of next-generation nuclear technologies.

Status and Future Prospects in Other Countries

Several countries are developing and testing Generation IV reactor technologies. For example, China has successfully started up a high-temperature gas-cooled modular pebble bed reactor, marking a significant milestone in the commercial deployment of Generation IV technology.

Sodium-cooled Fast Reactors (SFRs): SFRs are being utilized and developed worldwide for their ability to efficiently use nuclear fuel, reduce radioactive waste, and provide reliable power generation:

Russia:

- **BN-600:** Operational since 1980 at the Beloyarsk Nuclear Power Plant, with a capacity of 600 MWe. It has provided extensive operational data and experience.

- **BN-800:** Began operation in 2016 at the same plant, with a capacity of 800 MWe. This reactor uses mixed oxide (MOX) fuel, a blend of oxides of plutonium and uranium, and is designed to burn plutonium and minor actinides.

France:

- **Phénix:** Operated from 1973 to 2009, this 250 MWe prototype fast breeder reactor contributed significantly to the development of fast reactor technology and fuel cycle research.

- **Astrid:** The ASTRID nuclear project in France, aimed at developing a sodium-cooled fast reactor, was canceled. Initially planned as a 600 MWe prototype, it was later reduced to a 100-200 MWe research model before being halted in 2019. The cancellation was due to high costs, declining uranium prices, and safety concerns. France has shifted its focus to other energy strategies, delaying the deployment of fast reactors to later in the century.

Japan:

- **Monju:** A 280 MWe prototype SFR that faced operational challenges and was decommissioned in 2016. Despite this, Monju provided valuable insights into sodium handling and fast reactor technology.

Ongoing projects like ARC-100 and Natrium demonstrate the continued interest and investment in SFR technology for future global energy needs.

International and Industry Collaboration

International collaborations and public-private partnerships are critical for advancing nuclear technologies. The DOE and various industry partners are working together to address technical and regulatory challenges and promote the commercialization of advanced reactors.

Conclusion

Generation IV reactors represent the newest generation of advanced nuclear reactors, offering solutions to many of the challenges faced by previous generations. With their enhanced safety features, improved efficiency, and reduced waste production, these reactors have the potential to play a vital role in the global transition to clean energy. As research and development continue, and as prototype reactors prove their capabilities, Generation IV reactors are poised to become a cornerstone of the nuclear power industry in the coming decades. By exploring these cutting-edge technologies, we can envision a world powered by safe and sustainable nuclear energy, making significant strides in combating global warming.

References

1. Wikipedia. Generation IV reactor. https://en.wikipedia.org/wiki/Generation_IV_reactor

2. Cipiti, B., Sandia National Laboratories, National Technology and Engineering Solutions of Sandia LLC, & Honeywell International Inc. (2024). Gen-IV Proliferation Resistance & Physical Protection working group activities. In *ARSS Spring Program Review, INL* (pp. SAND2024-05271PE) [Report]. https://energy.sandia.gov/wp-content/uploads/2024/05/01-Cipiti-Gen-IV-Proliferation-Resistance-and-Physical-Protection-1.pdf

3. *3 Advanced reactor systems to watch by 2030.* (2021, April 12). Energy.gov. https://www.energy.gov/ne/articles/3-advanced-reactor-systems-watch-2030

4. Campbell, J. (n.d.). *The energy of innovation.* https://factsheets.inl.gov/Shared%20Documents/sodium-cooled-fast-reactor.pdf#search=Sodium%2DCooled%20Fast%20Reactor%20Fact%20Sheet

5. Fourth Generation Nuclear Reactors Take A Big Step Forward. Forbes, April 24, 2023. https://www.forbes.com/sites/rrapier/2023/04/24/fourth-generation-nuclear-reactors-take-a-big-step-forward/

6. Dumé, I. (2022, March 31). *Can thorium compete with uranium as a nuclear fuel?* Polytechnique Insights. https://www.polytechnique-insights.com/en/braincamps/energy/the-latest-technological-advances-in-nuclear-energy/can-thorium-compete-with-uranium-as-a-nuclear-fuel/

7. *Thorium's Long-Term Potential in nuclear energy: new IAEA analysis.* (2023, March 13). IAEA. https://www.iaea.org/newscenter/news/thoriums-long-term-potential-in-nuclear-energy-new-iaea-analysis

8. *Plans for new reactors worldwide - World Nuclear Association.* (2024, July 29). https://world-nuclear.org/information-library/current-and-future-generation/plans-for-new-reactors-worldwide

Chapter 9

Small Modular Nuclear Reactors: Pioneering the Future of Clean Energy

Small modular reactors (SMRs) have emerged as a revolutionary advancement in nuclear energy, offering substantial benefits over traditional large-scale reactors. These nuclear fission reactors, with a power output of up to 300 megawatts (MW) per module, are markedly smaller than conventional reactors, which often exceed 1,000 MW. Unlike their larger and more costly counterparts, SMRs are factory-built, designed for easy transportation to various locations by truck or ship.

The "modular" design means that major components can be assembled in a factory setting and then transported to the site for installation. Multiple units can be combined to achieve higher power outputs, reducing construction times and costs while maintaining high quality. This modularity not only enhances safety and flexibility in deployment but also allows for a more tailored energy solution. While not pocket-sized, their compact design provides unparalleled flexibility in meeting the diverse energy needs of communities and industries alike.

One of the most compelling advantages of SMRs is their ability to deliver electricity to smaller, isolated communities that lack access to traditional power grids. Beyond electricity generation, they are capable of powering desalination plants, supporting oil and gas operations, and providing process heat for industrial applications. Designed with safety as a paramount concern, SMRs often incorporate passive cooling systems that operate without external power during emergencies, and they are engineered to be more resilient to natural disasters such as earthquakes and floods.

As technological challenges are overcome, the cost of building SMRs is expected to drop below that of traditional nuclear power plants, making them an increasingly viable option for smaller communities and utilities.

Benefits of Small Nuclear Reactors

Given the promising future anticipated for SMRs and microreactors, let's summarize their key benefits:

- **Lower Capital Costs:** SMRs require a lower upfront capital investment compared to traditional reactors, making them accessible to a wider range of investors and utilities.

- **Quicker Deployment:** Factory-based construction accelerates deployment timelines, potentially reducing the build time from over a decade to just a few years.

- **Scalability:** SMRs can be scaled up or down to meet specific energy demands, making them suitable for both large and small grids.

- **Enhanced Safety Features:** SMRs incorporate advanced safety features, including passive systems that rely on natural physics—requiring no human intervention—to safely shut down and cool the reactor during abnormal conditions, thus minimizing the risk of accidents.

- **Reduced Environmental Impact:** SMRs produce low-carbon electricity, significantly contributing to the reduction of greenhouse gas emissions. They can also be integrated with renewable energy sources to ensure a stable and clean energy supply.

- **Siting Flexibility:** Due to their smaller size and lower cooling requirements, SMRs can be located in areas unsuitable for larger reactors, including remote regions and sites with limited water resources.

- **Minimal Land Use Needs:** SMRs require significantly less land than traditional nuclear plants and can even be placed on offshore barges to power onshore facilities. Compared to solar and wind farms, SMRs demand far less space—about 0.033 km²/GW versus 3.33 km²/GW for solar panels, where 1 GW equals 1 billion watts.

- **Versatile Applications:** Beyond generating electricity, SMRs are versatile tools in the clean energy transition, capable of providing industrial process heat, desalination, and hydrogen production.

Regulation of Small Modular Reactors

In the United States, the regulation of Small Modular Reactors (SMRs) is progressing as the Nuclear Regulatory Commission (NRC) works to establish a framework for their safe deployment. Here's an overview of the current status of SMR regulation in the U.S.:

1. **Licensing Framework:** The NRC has been adapting its regulatory framework to accommodate the unique features of SMRs. Unlike traditional large reactors, SMRs can be built modularly, in factories, and may be used in smaller or more remote locations. The NRC's existing regulations for large reactors are being updated to ensure they are applicable to SMRs, with particular attention to safety, security, and environmental impacts.

2. **Design Certification:** Several SMR designs are currently undergoing the NRC's design certification process, which is a key step in ensuring that a reactor meets all safety standards. Notably, NuScale Power became the first company to receive design approval for its SMR in 2020. This approval marks a significant milestone, as it paves the way for future deployment of NuScale's reactors in the U.S.

3. **Early Site Permits (ESPs) and Combined Operating Licenses (COLs):** The NRC offers companies the opportunity to apply for ESPs, which allow for the early approval of reactor sites before final design approval. Additionally, companies can apply for Combined Operating Licenses, which grant permission to

both construct and operate SMRs once they meet all regulatory requirements.

4. **Financial and Policy Support**: The U.S. government, through the Department of Energy (DOE), has been actively supporting SMR development. The DOE provides funding to companies developing SMR technologies and has partnered with private entities to advance demonstration projects. In 2020, the DOE selected a site at Idaho National Laboratory for the development of a NuScale SMR, aiming to have it operational by the late 2020s.

5. **Regulatory Challenges**: Despite progress, the NRC faces challenges in balancing the need for rigorous safety standards with the need to streamline regulatory processes for advanced technologies, especially for SMRs in the hope that many units can be produced with the same design under a single regulatory approval. SMRs introduce novel safety features, such as passive cooling systems, which require regulators to reconsider some traditional safety assessments used for larger reactors.

In summary, while the U.S. regulatory framework for SMRs is evolving, significant progress has been made with design certifications, site approvals, and federal support. The next few years are critical as SMRs move from the regulatory approval stage toward actual deployment.

Current Status of Development

Globally, SMR development is gaining impressive momentum, with over 70 designs in various stages of development. The industry's growth is reflected in the global SMR project pipeline, which has reached approximately 22 GW, representing a potential investment of $176 billion—a 65% increase since 2021. Countries leading SMR development include the United States, United Kingdom, China, Canada, France, South Korea, and Russia. The United States leads with nearly 4 GW in planned SMR capacity, followed by Poland and Canada, each with around 2 GW.

Notable Projects and Developments

SMR designs span a variety of reactor types, including light water reactors, high-temperature gas-cooled reactors, liquid metal-cooled fast reactors, and molten salt reactors. Some of the key projects include:

1. **China:** The ACP100 SMR, also known as Linglong One, is progressing with its outer containment dome installed as of February 2024. The project aims for completion by the end of 2026. The HTR-PM demonstration plant was connected to the grid in December 2021, and in 2023, China activated the world's first commercial 210 MW SMR.

2. **Russia:** The Akademik Lomonosov floating nuclear power plant, with two KLT-40S reactors, has been operational since May 2020. Russia is also developing the land-based RITM-200N reactor, expected to be completed by 2028.

3. **NuScale Power (U.S.):** NuScale Power, the leading developer of small modular reactor technology, has experienced a mix of challenges and progress in recent months. Despite the cancellation of its first planned commercial project in Idaho due to rising costs, the company has shown resilience and potential for growth in 2024. NuScale remains the only SMR developer with design certification from the U.S. Nuclear Regulatory Commission, giving it a competitive edge in the market. They have made strides in international partnerships, and there is also growing interest from data centers and manufacturers seeking reliable, emissions-free electricity, which aligns well with NuScale's technology. While challenges persist, particularly in proving the economic viability of its technology, NuScale remains optimistic about its future in the expanding market for clean energy solutions.

4. **X-energy (U.S.):** X-energy is planning a four-unit facility at Dow Chemical's site in Texas, with construction expected to begin in 2026.

5. **Oaklo (U.S.):** Backed by Sam Altman, Oaklo is advancing its first small modular reactor project at Idaho National Laboratory. The company is progressing through regulatory processes with the U.S. Nuclear Regulatory Commission and has received a site use permit from the Department of Energy. Oklo has partnered with Wyoming Hyperscale to supply power to

a data center and has secured over $300 million in funding through a merger with AltC Acquisition Corp. These highlight Oklo's focus on rapid construction, regulatory compliance, and strategic partnerships for clean energy solutions.

6. **Rolls-Royce SMR (UK):** Rolls-Royce is developing a 470 MW SMR, with ongoing regulatory assessments for deployment in Poland.

7. **GE Hitachi (U.S.):** The BWRX-300 reactor design is gaining traction in Poland, with plans approved for up to 24 units across six sites.

8. **South Korea:** In July 2023, South Korea launched a $305 million project to develop a national SMR by 2028, forming an alliance with 42 entities.

9. **France:** The French government increased funding for SMR development, with EDF's Nuward project receiving additional support for design and safety demonstrations.

10. **Copenhagen Atomics and Paul Scherrer Institute**: These Danish and Swiss companies have entered into a collaboration to conduct a thorium molten salt critical experiment by 2026. This experiment aims to validate the technology and gather data for commercial deployment. Copenhagen Atomics is developing a containerized molten salt reactor that uses thorium and low-enriched uranium fluoride salt, with plans for commercial reactors by 2028.

11. **Thorizon and Naarea**: Thorizon of the Netherlands and Naarea of France have formed a strategic partnership to advance molten salt reactor (MSR) technology in Europe. Thorizon is working on a 250 MWt/100 MWe MSR, aiming for a pilot reactor by 2035, while Naarea is developing an ultra-compact fast neutron reactor. The collaboration focuses on pooling resources for safety demonstrations and developing shared facilities.

12. **Argentina:** The CAREM25 SMR is under construction, with grid connection expected by 2026. The project has faced delays from its original timeline.

Challenges and Considerations

Despite the promising outlook, several challenges remain:

- **Regulatory Hurdles:** Licensing and regulatory approval processes for new reactor designs are complex, requiring substantial governmental support and international collaboration.

- **Economic Viability:** Achieving cost competitiveness with other energy sources remains challenging. The industry must develop a global market and realize economies of scale through mass production.

- **Fuel Supply:** Some SMR designs require high-assay low-enriched uranium, necessitating the establishment of dedicated production capabilities.

- **Realistic Deployment Timelines:** While optimistic projections a decade ago suggested 65-75 GW of SMR capacity by 2035, more recent estimates suggest closer to 6 GW by that time. Significant deployments are now expected around 2030-2035, with advanced Gen IV SMRs potentially facing greater delays.

- **Supply Chain and Infrastructure:** Developing robust supply chains and necessary infrastructure for SMR deployment remains a critical challenge.

Yet, the growing interest from utilities and industrial end-users—such as oil and gas extractors, petrochemical processors, and data centers—signals a bright future for SMRs in the clean energy transition.

Microreactors: Smaller than Small

Microreactors are ultra-compact nuclear reactors designed to generate up to 20 megawatts of thermal energy, offering a versatile solution for both electricity generation and industrial applications. Unlike traditional reactors and even small modular reactors, microreactors are significantly smaller, portable, and adaptable, often serving as a cleaner alternative to diesel generators. These reactors can be fully assembled in factories and transported by truck, rail, water, or even air. Designed to operate independently or within a microgrid, they provide resilient, non-carbon-emitting power in challenging environments.

The concept of microreactors dates back to the U.S. Navy's nuclear submarine project in the 1950s. The first nuclear submarine, the USS *Nautilus*, launched in 1955, was powered by a small reactor generating about 10 megawatts. In the 1960s, the U.S. military deployed portable nuclear reactors to remote locations, including Greenland, Antarctica, and Alaska. These early versions demonstrated the feasibility of micro-reactors but were eventually phased out as military priorities shifted.

Today's microreactors present compelling advantages in the fight against global warming. Their compact size and modular design enable rapid deployment and scalability, making them an ideal solution for meeting the urgent demand for clean energy. By generating low-carbon electricity, microreactors contribute significantly to reducing greenhouse gas emissions. They also enhance energy security, providing reliable, round-the-clock power in remote or disaster-stricken areas. Some designs can even operate for up to 10 years without needing refueling, offering long-term, low-maintenance energy solutions.

Microreactors can also be integrated with renewable energy sources to form hybrid energy systems, ensuring a stable and sustainable energy supply. Their applications extend beyond electricity generation to include industrial heat, desalination, and hydrogen production, making them valuable assets in the transition to clean energy.

However, despite their promise, microreactors face challenges such as regulatory hurdles, economic viability, and public perception.

Current Developments

Several countries and organizations are advancing microreactor technology. In the U.S., the Department of Energy (DOE) supports various projects, including the MARVEL reactor at Idaho National Laboratory which aims to demonstrate the integration of microreactors with other energy sources. The U.S. Department of Defense is also exploring military applications through initiatives like Project Pele—a high-temperature, gas-cooled microreactor scheduled for testing in 2025.

The U.S. government is also investing in small modular reactors (SMRs), providing up to $900 million in funding through the Bipartisan Infrastructure Law. This funding supports SMR development as a cost-effective, scalable alternative to traditional nuclear plants, vital for achieving net-zero emissions by 2050. Companies like TerraPower, Oklo, and NuScale Power are at the forefront of SMR innovation, with tech giants like Microsoft and Amazon Web Services exploring nuclear energy to meet their sustainability goals.

James Walker, CEO of Nano Nuclear Energy, emphasizes that microreactors, expected to launch commercially by 2031 at costs as low as $20 million, could meet the growing energy demands of AI-driven data centers. These reactors will be portable, easily installed, and designed with advanced safety features to prevent overheating or meltdowns. While regulatory approval remains a challenge, initiatives like the U.S. Department of Defense's Project Pele could streamline the process. Walker also addresses concerns about nuclear waste, pointing out that the amount generated by microreactors is minimal and easily managed, positioning them as the safest and cleanest energy option available.

Other private companies are making strides as well. Westinghouse is developing the eVinci microreactor, capable of generating up to 5 megawatts of electricity and operating for eight years without refueling. Ultra Safe Nuclear Corporation's Pylon microreactor, designed for both terrestrial and space applications, and Radiant's Kaleidos microreactor, a 1.2-megawatt portable reactor, aim to replace diesel generators in remote locations. These advances demonstrate the broad potential of microreactors to address diverse energy needs.

The Future of SMRs in Clean Energy Production

Small Modular Reactors (SMRs) represent a pivotal advancement in the future of clean energy. Their factory-based manufacturing process not only ensures better standardization and quality control but also streamlines regulatory approval, potentially slashing deployment timelines from over a decade to just a few years. This efficiency in production and regulatory approval is vital for accelerating our transition to clean, sustainable energy systems.

SMRs provide reliable, low-carbon power in a flexible and scalable manner, making them an ideal choice for countries committed to reducing their carbon footprints. Their adaptability to diverse energy needs and environments positions them as critical players in the global effort to achieve net-zero emissions and combat climate change. As technological advancements continue and regulatory frameworks evolve, SMRs are set to play an increasingly crucial role in the global energy landscape.

Similarly, microreactors, like their larger SMR counterparts, stand as a significant innovation in nuclear energy. They offer flexible, scalable, and low-carbon solutions to meet the growing demand for clean energy. With ongoing technological advancements and the development of supportive regulatory frameworks, microreactors are well-positioned to contribute meaningfully to the global push for net-zero emissions and the fight against climate change.

SMRs bring together a powerful combination of cost-effectiveness, safety, and flexibility, making them perfectly suited to meet the ever-increasing demand for clean, reliable energy. As their development progresses and deployment scales up, SMRs could very well become a cornerstone of the world's clean energy future, ensuring a sustainable and resilient energy supply for generations to come.

References

1. International Atomic Energy Agency (IAEA). Small modular reactors. (n.d.). https://www.iaea.org/topics/small-modular-reactors

2. *NRC certifies first U.S. small modular reactor design.* (2023, January 20). Energy.gov. https://www.energy.gov/ne/articles/nrc-certifies-first-us-small-modular-reactor-design

3. Idaho National Laboratory. (2024, April 22). Advanced small modular reactors. https://inl.gov/trending-topics/small-modular-reactors

4. Office of Nuclear Energy. 4 Key benefits of advanced small modular reactors. (n.d.). https://www.energy.gov/ne/articles/4-key-benefits-advanced-small-modular-reactors

5. Nuclear Energy Agency (NEA). Small modular reactors: challenges and opportunities. (2021). https://www.oecd-nea.org/

jcms/pl_57979/small-modular-reactors-challenges-and-op-portunities?details=true

6. Cleantech Group. Gilani, Z. (2023, Dec. 22). Will small modular reactors surpass regulatory and supply chain hurdles to fill the need for stable, baseload power? https://www.cleantech.com/will-small-modular-reactors-surpass-regulatory-and-supply-chain-hurdles-to-fill-the-need-for-stable-baseload-power/

7. Idaho National Laboratory. (2024, April 22). Microreactors. https://inl.gov/trending-topics/microreactors/

8. New MARVEL project aims to supercharge microreactor deployment. (April 13, 2021). Energy.gov. https://www.energy.gov/ne/articles/new-marvel-project-aims-supercharge-microreactor-deployment

9. Vox. Irfan, U. (2023, May 1). Smaller, cheaper, safer: The next generation of nuclear power, explained. https://www.vox.com/science/23702686/nuclear-power-small-modular-reactor-energy-climate-change

10. Microgrid Knowledge. Rod Walton. (June 20,2024,). Small Nuclear's big moment: DOE funding $900M for SMR projects. https://www.microgridknowledge.com/generation-fuels/article/55090038/small-nuclears-big-moment-doe-funding-900m-for-smr-projects

11. *Akademik Lomonosov floating Nuclear Co-Generation Plant, Kamchatka.* (2021, November 5). Power Technology. https://www.power-technology.com/projects/akademik-lomonosov-nuclear-co-generation-russia/

12. *SMR Technology trends worldwide. (2024, June 20). Enerdata.* https://www.enerdata.net/publications/executive-briefing/smr-world-trends.html

13. Defense Logistics Agency. (Sept. 5, 2023). Micro-reactor pilot program reaches major milestone. https://www.dla.mil/About-DLA/News/News-Article-View/Article/3515062/micro-reactor pilot-program-reaches-major-milestone/

14. Nuclear Newswire. (2023, Aug. 28). Project Pele in context: An update on the DOD's microreactor plans. https://www.ans.org/news/article-5308/project-pele-in-context-an-update-on-the-dods-microreactor-plans/

15. Micro-Reactor. (n.d.). https://www.rolls-royce.com/innovation/novel-nuclear/micro-reactor.aspx
16. Williams, W. (2024, September 11). Micro nuclear reactors could cost as little as $20 million and launch by 2031 — but will it be enough for data center... *TechRadar.* https://www.techradar.com/pro/micro-nuclear-reactors-could-cost-as-little-as-20-million-and-launch-by-2031-but-will-it-be-enough-for-data-center-operators-and-the-ai-industry
17. *GE Hitachi, Holtec, Rolls-Royce SMR and Westinghouse enter UK SMR negotiations.* (2024, September 26). World Nuclear News. https://www.world-nuclear-news.org/articles/ge-hitachi-holtec-rolls-royce-smr-and-westinghouse-enter-uk-smr-negotiations

Chapter 10

Nuclear Fusion: Unlocking the Power of the Stars on Earth

In this chapter, we embark on an exhilarating journey into the realm of nuclear fusion—the very process that powers the stars—and explore its transformative potential as a game-changing energy source. Nuclear fusion promises clean, abundant, and sustainable energy without the drawbacks associated with traditional nuclear fission. Join me as we delve into the fascinating world of nuclear fusion and its pivotal role in combating climate change.

Understanding Nuclear Fusion

Nuclear fusion is a complex yet fascinating process that involves merging two light atomic nuclei to form a heavier nucleus, releasing a tremendous amount of energy. This is the very process that powers the Sun and other stars, positioning it as a potential source of clean, limitless energy here on Earth.

How Fusion Works

Nuclear fusion occurs when two light atomic nuclei, such as isotopes of hydrogen (deuterium and tritium), collide under extreme conditions of temperature and pressure. The fusion process can be summarized as follows:

- **High Temperatures:** To achieve fusion, the nuclei must be heated to incredibly high temperatures, often exceeding 100 million degrees Celsius. At these extreme temperatures, the hydrogen isotopes become a plasma—a state of matter where electrons are stripped away from nuclei.

- **Overcoming Electrostatic Repulsion:** Nuclei are positively charged and naturally repel each other due to electrostatic forces. For fusion to occur, the kinetic energy of the nuclei must be sufficient to overcome this repulsion. This is achieved through the high temperatures, which accelerate the particles to speeds that increase their energy.

- **Strong Nuclear Force:** Once the nuclei are close enough, the strong nuclear force—which is far more powerful than electrostatic forces at very short distances—takes over, binding the nuclei together and resulting in fusion.

- **Energy Release:** The fusion of deuterium and tritium produces a helium nucleus and a high-energy neutron, releasing a very large amount of energy due to the difference in nuclear binding energy between the starting deuterium and tritium and the fusion products.

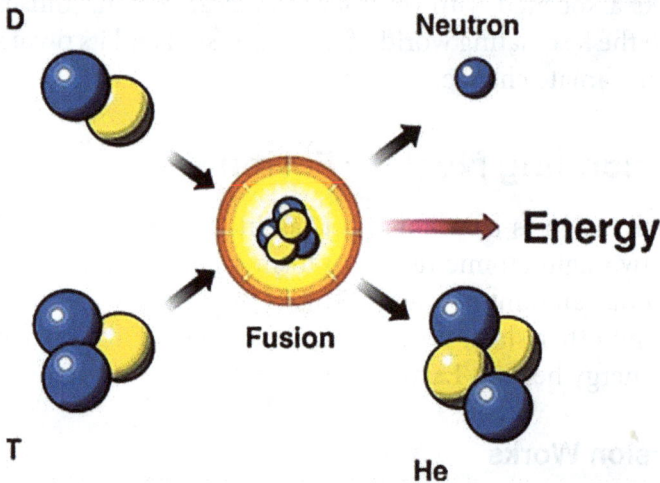

NUCLEAR FUSION

The Promise and Challenges of Fusion Energy

Nuclear fusion holds the potential to revolutionize our energy landscape, offering a nearly limitless source of clean, sustainable energy. Unlike traditional nuclear fission, fusion produces minimal radioactive waste, has no carbon emissions, and taps into abundant fuel sources—making it a compelling solution for our growing energy needs and the fight against climate change.

Advantages of Fusion Energy

Abundant Fuel: The primary fuels for fusion—two isotopes of hydrogen, deuterium and tritium—are plentiful. Deuterium can be extracted from seawater, and tritium can be bred from lithium, both of which are abundant resources.

Minimal Waste: Fusion generates far less radioactive waste compared to nuclear fission, and the waste it does produce has a much shorter half-life, simplifying waste management.

No Carbon Emissions: Fusion does not emit carbon dioxide, making it a powerful tool in reducing greenhouse gases and combating climate change.

Safety: Fusion reactions do not produce highly radioactive byproducts, eliminating the risks associated with nuclear fission, such as meltdowns or long-term waste storage. The primary byproduct of fusion is helium—an inert, non-toxic gas.

Environmental Impact: Fusion produces no greenhouse gases or long-lived radioactive waste, making it an environmentally friendly energy source.

Overcoming Fusion Challenges

Despite its immense potential, harnessing nuclear fusion for practical energy production presents formidable challenges. The key obstacles revolve around plasma confinement, heat management, the tritium fuel cycle, and the complex engineering required to scale fusion power to a commercial level.

- **Plasma Confinement:** One of the fundamental challenges is containing the super-heated plasma for extended periods. Developing advanced materials and technologies that can withstand these extreme conditions is crucial. Current research focuses on two primary methods of plasma confinement: magnetic confinement and inertial confinement.

- **Magnetic Confinement:** This method uses strong magnetic fields to contain the hot plasma. The most common device is the tokamak, a doughnut-shaped chamber where plasma is confined in a circular path. Advances include the ITER project, which aims to produce sustained fusion reactions. Another approach is the stellarator, which twists the magnetic field into a complex shape to enhance stability.

- **Inertial Confinement:** This method involves compressing small fuel pellets with intense laser or ion beams to achieve the necessary temperatures and pressures for fusion. The National Ignition Facility (NIF) in the U.S. is a leading project, having recently achieved net energy gain in an experiment.

- **Emerging Techniques:** Researchers are also exploring alternative methods, such as magnetic target fusion, which combines elements of both magnetic and inertial confinement. Other innovations include advanced magnetic confinement devices like spherical tokamaks and high-beta configurations.

- **Heat Management:** Efficiently extracting and utilizing the intense heat generated by fusion reactions remains a significant engineering challenge. This is crucial not only for energy production but also for maintaining the structural integrity of reactor components.

- **Tritium Fuel Cycle:** Establishing sustainable methods for producing and recycling tritium, a key fusion fuel, is essential for long-term viability. Tritium, being radioactive and rare in nature, requires careful handling and efficient production methods to support continuous fusion reactions.

- **Engineering and Systems Integration:** Creating practical fusion power plants that can reliably supply energy on a commercial scale requires solving many complex engineering problems. This includes developing materials that can endure extreme environments and integrating various systems to ensure consistent and safe operation.

Fusion's Transformative Potential

Despite these challenges, the potential of nuclear fusion as an energy source is truly staggering. Fusion reactions release many times more energy than traditional chemical reactions, such as the combustion of fossil fuels. The abundant fuel sources and minimal environmental impact position fusion as a cornerstone of a sustainable energy future.

The Road Ahead for Fusion Energy

Interest and investment in fusion energy are surging, driven by the promise of a transformative solution to the world's energy challenges. In 2022 alone, private fusion companies reported $2.83 billion in new investments, underscoring the growing confidence in fusion's potential. With ongoing research, technological advancements, and increasing global support, fusion power could revolutionize our energy landscape in the coming decades.

The 21st century has witnessed a renewed momentum in fusion research, fueled by technological breakthroughs and the urgent need to combat climate change. As promising methods and recent breakthroughs bring us closer to harnessing this clean, abundant energy source for commercial use, several private companies have entered the fusion race. Companies like Commonwealth Fusion Systems, Tokamak Energy, and TAE Technologies are leading the charge with innovative reactor designs and accelerated timelines, backed by significant investment.

However, while nuclear fusion holds immense promise, significant scientific and engineering challenges remain. Overcoming these obstacles will require sustained effort, but the potential rewards—a limitless, clean energy supply for many generations—make this pursuit essential.

International Collaboration in Fusion Research

Recognizing the global significance of fusion energy, countries have joined forces to accelerate progress and share knowledge. The International Thermonuclear Experimental Reactor (ITER) project exemplifies this collaboration, bringing together 35 nations to demonstrate the scientific and technical feasibility of fusion power. Other research initiatives, such as the National Ignition Facility in the United States and the Joint European Torus in the United Kingdom, contribute to advancing fusion science through their pioneering efforts.

Key Recent Developments

- **Inertial Confinement Breakthrough:** In December 2022, researchers at the National Ignition Facility in California achieved a major milestone by demonstrating net energy gain in an inertial confinement fusion experiment for the first time. While still far from practical energy production, this break-through represents a significant leap forward in fusion science.

- **ITER:** The ITER project in France, currently under construction, represents decades of international collaboration in fusion research. When completed, ITER will be the world's largest to-kamak, designed to produce 500 megawatts of fusion power for extended periods. Despite facing delays and cost overruns, ITER remains the cornerstone of the global fusion effort.

- **High-Temperature Superconductors:** The development of high-temperature superconducting magnets has opened new possibilities for more compact and efficient fusion reactors. These magnets can generate stronger magnetic fields, poten-tially enabling smaller and less expensive fusion devices.

- **SPARC:** A promising new experimental device, SPARC (Soon-est or Smallest Privately funded Affordable Robust Compact reactor), is being developed by scientists at MIT and Common-wealth Fusion Systems. SPARC aims to use high-temperature superconducting magnets to confine plasma and could gen-erate up to 10 times more energy than is input. If successful, SPARC would be the first device to achieve a burning plasma, where the heat from fusion reactions sustains the process without additional energy input.

- **France's Tokamak Record**: A fusion reactor in southern France recently set a new record by creating and sustaining plasma that carried 15% more energy and was twice as dense as previous attempts. The reactor maintained its plasma for over 17 minutes at an astonishing 158,000,000°F—five times hotter than the Sun. This achievement brings us closer to practical nuclear fusion power, though significant challenges, particularly in extending plasma life, remain.

Future Outlook and Challenges

Despite these remarkable advancements, achieving commercial fusion power will require overcoming significant challenges. Maintaining plasma stability and confinement over extended periods remains a formidable task due to inherent instabilities. Developing materials capable of withstanding the intense heat and neutron bombardment inside a fusion reactor is crucial for long-term operation. Additionally, establishing efficient tritium breeding systems is essential, given the rarity of tritium in nature. Converting the immense heat produced by fusion reactions into electricity presents further engineering challenges, and reducing costs and simplifying designs are critical for commercial viability.

While most experts agree that commercially viable fusion power is unlikely before 2050, even in optimistic scenarios, the immense potential rewards drive continued research and development. Virtually limitless clean power with minimal environmental impact remains a tantalizing goal. As climate change intensifies and the demand for carbon-free energy grows more urgent, sustained investment in fusion research is crucial for realizing this long-held dream.

Nuclear fusion, the process that powers the stars, holds immense promise as a future energy source on Earth. Its potential to provide abundant, clean, and sustainable energy positions fusion to play a pivotal role in combating climate change and meeting global energy demands. While significant challenges persist, international collaboration and ongoing research are bringing us closer to the realization of fusion power. The future of fusion energy is bright, with the potential to reshape electricity generation, address the pressing challenges of our time, and provide clean energy far into our future.

Author's Note: Many years ago, while I was finishing high school and as an undergraduate at the University of California at Berkeley, I worked summers at the Lawrence Radiation Laboratory in Livermore, California, which is now home to the National Ignition Facility. Nuclear fusion was being studied back then with long-term hopes of being able to harness it as a power source. Here we are, decades later, closer to that goal, but with a long way yet to go.

Cold Fusion: Real Possibility or Imagined Dream?

Cold fusion, a hypothesized nuclear reaction occurring at or near room temperature, contrasts sharply with the extreme conditions required for traditional fusion. In 1989, scientists Martin Fleischmann and Stanley Pons claimed to have achieved cold fusion through the electrolysis of heavy water using a palladium electrode, reportedly producing excess heat and nuclear byproducts. However, their findings faced widespread skepticism due to failed replication attempts and a lack of theoretical support. While mainstream science largely dismissed cold fusion as pathological science, a small group of researchers continues to explore it under the term low-energy nuclear reactions (LENR).

Scientific Controversy and Challenges

The announcement by Fleischmann and Pons sparked excitement and skepticism, centered on key issues:

- **Reproducibility**: Failed attempts to replicate the original experiment cast doubt on its validity.

- **Theoretical Basis**: The absence of a widely accepted theory explaining cold fusion at room temperature left scientists unconvinced.

- **Expected Byproducts**: The lack of typical nuclear byproducts, such as gamma rays and neutrons, fueled skepticism.

- **Experimental Flaws**: Potential errors and misinterpretations in the experimental setup undermined the initial findings.

As a result, cold fusion was largely relegated to the fringes of scientific inquiry. However, its potential implications have kept a dedicated group of researchers investigating LENR.

Current Status and Future Prospects

Interest in cold fusion persists on a smaller scale as researchers explore LENR, seeking evidence or a theoretical framework to validate the phenomenon. The field remains controversial, with inconsistent results and no widely accepted theory behind it.

Recent developments, such as the Department of Energy's Advanced Research Projects Agency–Energy (ARPA-E) announcement of $10 million in funding for eight projects designed to determine whether low-energy nuclear reactions (LENR)—historically and sometimes disparagingly known as "cold fusion"—could someday be a carbon-free energy source. ARPA-E intends the funding to "break the stalemate" and determine if LENR holds any merit for future energy research.

The selected projects involve universities, a national laboratory, and small businesses, focusing on various aspects of LENR, including:

- **University of Michigan**
- **Texas Tech University**
- **Lawrence Berkeley National Laboratory**
- **Massachusetts Institute of Technology (MIT)**
- **Stanford University**
- **Energetics Technology Center**
- **Amphionic**

The scientific community remains divided over cold fusion's possibility, much less feasibility as a practical energy source, and cold fusion research continues to face significant criticisms:

- **Lack of Reproducibility**: Many experiments have failed to replicate the original findings.

- **Absence of Expected Byproducts**: Anticipated nuclear byproducts have not been observed.

- **Theoretical Challenges**: No widely accepted theory explains cold fusion at room temperature.

- **Experimental Flaws**: Errors in experimental setups have been identified.

- **Pathological Science**: Cold fusion is often seen as inconsistent with established scientific principles.

- **Premature Publicity**: The initial announcement led to publicity and investment before sufficient peer review.

Despite these criticisms, some researchers remain intrigued by cold fusion's possibilities. While initial claims have not been universally accepted or reliably replicated, the idea of room-temperature fusion continues to captivate some in the scientific community. The ongoing exploration of cold fusion highlights the dynamic nature of scientific inquiry and the importance of rigorous experimentation and open dialogue. As it stands, cold fusion remains more of a scientific question mark than a viable solution to the world's energy needs.

Author's Note: While I do think one day scientists and engineers will achieve high-temperature sustained fusion and even fusion power plants, cold fusion breaks the laws of physics and will simply never be possible.

References

1. Encyclopedia Britannica. Conn, R. W. (May 21,2024). Nuclear fusion | Development, Processes, Equations, & Facts. https://www.britannica.com/science/nuclear-fusion/History-of-fusion-energy-research
2. IAEA. Safety in fusion. (May 2021). https://www.iaea.org/bulletin/safety-in-fusion
3. The Economist. Is nuclear fusion the future of clean energy? (Jan. 2024). https://www.youtube.com/watch?v=WuOBR7t-1mG8

4. World Nuclear Association. CURRENT AND FUTURE GEN-ERATION: Nuclear Fusion Power. (Dec.22, 2022.). https://world-nuclear.org/information-library/current-and-future-generation/nuclear-fusion-power

5. CNN World. Paddison, L. (Dec. 21, 2023). Scientists successfully replicate historic nuclear fusion breakthrough three times. https://www.cnn.com/2023/12/20/climate/nuclear-fusion-energy-breakthrough-replicate-climate/index.html

6. EUROfusion. (June 7, 2024,). History of Fusion - EUROfusion. https://euro-fusion.org/fusion/history-of-fusion/

7. ITER. What will ITER do? (n.d.). https://www.iter.org/sci/Goals

8. *Nuclear fusion reactor for cleaner energy.* (2024.). https://www.msn.com/en-us/money/companies/nuclear-fusion-reactor-for-cleaner-energy/vi-BB1iM6sc?ocid=socialshare#details

9. *Record-breaking nuclear fusion advance.* (2024, July). https://www.msn.com/en-us/money/companies/record-breaking-nuclear-fusion-advance/vi-BB1maPjs?ocid=msedgntp&pc=HCTS&cvid=0cb1f76583a1473eb-77ba9123cb1f4fc&ei=27#details

10. Understanding Science. (2022, August 12). *Cold fusion - Understanding Science.* Understanding Science - How Science REALLY Works. . . https://undsci.berkeley.edu/cold-fusion-a-case-study-for-scientific-behavior/

11. *ARPA-E picks eight teams to prove—or debunk—low-energy nuclear reactions.* (2023, February 23). - ANS / Nuclear Newswire. https://www.ans.org/news/article-4769/arpae-picks-eight-teams-to-proveor-debunklowenergy-nuclear-reactions/

Chapter 11

The Economic Viability of Nuclear Power: Costs and Benefits

The economic aspects of nuclear power are crucial to understanding its viability as a long-term energy solution. As we explore its round-the-clock reliability, low carbon emissions, high initial costs, safety concerns, and long-term waste management, it's essential to assess how nuclear power compares with other energy sources. Join us as we delve into the economics of nuclear power and its role in our transition to a sustainable energy future.

Understanding the Costs of Nuclear Power

The economic viability of nuclear power is multifaceted, involving upfront construction costs, operational expenses, and long-term waste management. While the initial investment is substantial, factors such as operational efficiency, low carbon emissions, and energy security strengthen the economic case for nuclear power. Government policies and incentives also play a pivotal role in enhancing its competitiveness. By thoroughly evaluating these costs and benefits, we can better gauge the economic feasibility of nuclear power as a sustainable and low-carbon energy solution.

Upfront Capital Costs

The upfront capital costs of building nuclear power plants are among the most significant economic challenges due to the complexity of their design and construction. These costs encompass engineering, procurement, construction, and regulatory compliance. Factors such as the availability of skilled labor, materials, and stringent safety requirements further influence the overall cost of nuclear projects.

Operational Costs and Lifecycle Analysis

Beyond construction, operational costs are crucial in determining the economic viability of nuclear power. These include fuel procurement, routine maintenance, staffing, and waste management. A comprehensive lifecycle analysis, considering costs over the entire lifespan of a nuclear plant, is essential to assess the long-term economic competitiveness of nuclear power compared to other energy sources.

Levelized Cost of Electricity

The levelized cost of electricity (LCOE) is a key metric for comparing the economic competitiveness of different energy sources. LCOE factors in the total costs of building, operating, and maintaining a power plant over its expected lifetime, alongside the electricity produced. While LCOE estimates for nuclear power vary based on location, design, and financing, advanced designs are expected to be more competitive with other low-carbon energy sources like wind and solar over their lifetimes.

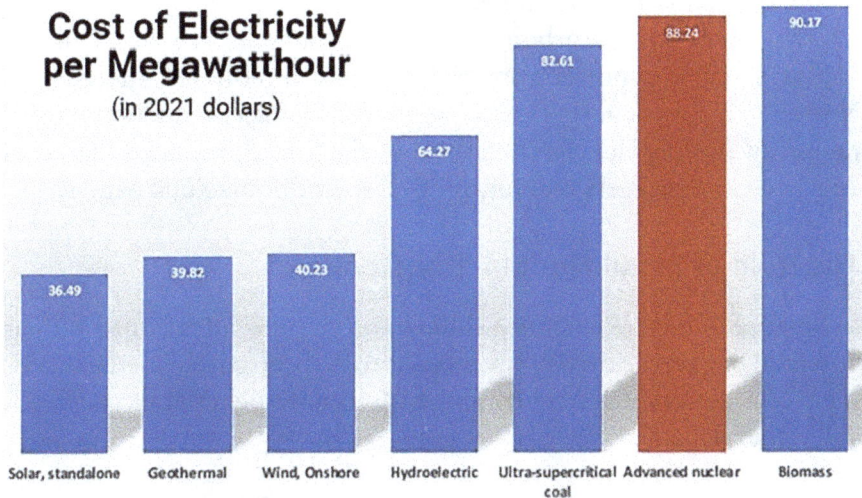

Cost of Electricity per Megawatthour
(in 2021 dollars)

Solar, standalone	Geothermal	Wind, Onshore	Hydroelectric	Ultra-supercritical coal	Advanced nuclear	Biomass
36.49	39.82	40.23	64.27	82.61	88.24	90.17

According to the World Nuclear Association, capital costs account for at least 60% of the LCOE for nuclear plants. Interest charges during lengthy construction periods further escalate costs. However, in countries with continuous nuclear development programs, like South Korea, capital costs have been better contained. A recent project in the United States, the expansion of Plant Vogtle in Georgia, has faced substantial cost overruns and delays, with the project ultimately costing nearly $35 billion—more than double the initial projections. Such escalations highlight the challenges of financing new nuclear projects.

External Costs and Benefits

Evaluating the economic viability of nuclear power also requires considering external costs and benefits. External costs include environmental impacts, waste management, and decommissioning expenses. On the other hand, nuclear power offers significant benefits, such as low carbon emissions, energy security, and job creation. Quantifying and comparing these external factors is crucial for a comprehensive economic assessment.

Government Policies and Incentives

Government policies and incentives are instrumental in shaping the economic landscape of nuclear power. Subsidies, tax credits, and loan guarantees can offset upfront costs and improve the competitiveness of nuclear projects. Additionally, carbon pricing mechanisms that internalize the cost of carbon emissions further enhance the economic appeal of nuclear power compared to fossil fuel-based alternatives. Notably, nuclear energy is the most land-efficient energy source, requiring 50 times less land per unit of electricity produced than coal and 18 to 27 times less than on-ground solar photovoltaic panels.

The Role of Small Modular Reactors

Small Modular Reactors (SMRs) are gaining attention for their potential to provide cost-effective, low-carbon baseload. However, their cost competitiveness compared to renewable energy sources is still under debate:

LCOE: Estimates for SMRs range from \$45-\$65/MWh, with some sources suggesting higher costs around \$94.62/MWh. This variability reflects uncertainties in deploying SMRs at scale.

Long-Term Cost Effectiveness: Over a 60-year period, SMRs are expected to be cost-effective compared to other baseload power sources, including renewables with battery storage, considering their consistent and reliable power output.

Initial Capital Costs: While SMRs require lower upfront capital investment than traditional large-scale reactors, they remain more expensive than renewable sources like wind and solar.

Comparing Renewable Energy (Wind and Solar)

Cost Declines: Solar and wind energy costs have decreased dramatically, with solar dropping by 90% and wind by 70% in the last decade. This trend is expected to continue with technological advancements and economies of scale.

Battery Storage: When combined with battery storage for reliability, renewable energy costs increase significantly. However, the declining costs of battery technology will further enhance the economic viability of renewables.

Efficiency Benefits of Renewable Energy

Renewable energy sources offer distinct efficiency benefits:

Resource Utilization: Wind and solar energy harness abundant and free natural resources, with efficiency primarily dependent on geographic and climatic conditions.

Technological Advancements: Advances in photovoltaic technology and wind turbine design have significantly improved the efficiency of these systems, making them more competitive with traditional and nuclear power sources.

Integration with Storage: Pairing renewable energy with battery storage can provide a stable and reliable power supply, though this integration may reduce overall efficiency due to energy losses in storage and retrieval processes.

Insights from The Lazard 2024 LCOE+ Report

The Lazard 2024 Levelized Cost of Energy (LCOE+) report summarizes and offers a comprehensive comparison of the LCOE for various renewable energy sources and next-generation nuclear power plants, including small modular reactors (SMRs). In this section, we'll explore key findings from the report.

It's important to recognize that LCOE calculations can vary based on assumptions about financing costs, capacity factors, and plant lifespans. Nuclear plants typically have very long operational lives, which can improve their long-term economics.

The report illustrates that renewable energy sources like utility-scale solar photovoltaic (PV) panels and onshore wind alone have significantly lower LCOEs compared to traditional nuclear power. However, the inclusion of attached storage substantially raises the overall cost of solar and wind compared to their standalone installations.

The SMRs present a competitive LCOE range, though they generally remain higher than the cheapest renewable options. Here's a simplified representation of the LCOE comparisons:

Energy Source	LCOE Range ($/MWh)
Utility-Scale Solar Photovoltaic	29 - 92
Utility-Scale Solar Photovoltaic with Attached Storage	60 - 210
Onshore Wind	27 - 73
Onshore Wind with Attached Storage	45 - 133
Offshore Wind	72 - 140
Next-Generation Large Nuclear Plants	131 - 204
Small Modular Reactors	60 - 110

The Lazard data highlights the cost competitiveness among clean energy sources while showing that SMRs could offer a more affordable nuclear option compared to large nuclear plants, though large nuclear plants are still generally more expensive than solar and onshore wind.

Conclusion

The economic viability of nuclear power remains a complex and debated issue within the energy sector. While nuclear plants have high upfront capital costs, they provide reliable baseload power with low operating costs over decades-long lifespans. However, several factors complicate the economic picture.

The LCOE of nuclear power varies widely depending on the specific assumptions and projects under consideration. While some estimates suggest that nuclear power can be competitive with other baseload energy sources, others indicate that its costs are significantly higher compared to alternatives like natural gas or renewables. However, when the costs of storage are factored into the LCOEs for solar and wind, nuclear power becomes more competitive. This variability underscores the complexity of nuclear economics and highlights the significant differences between individual plants and markets.

Construction costs and timelines are major challenges for nuclear projects. Many recent projects have faced severe cost overruns and delays, increasing financial risks and making nuclear less attractive to investors compared to other energy sources. However, standardized designs and improved project management may help address these issues in future projects.

Operating costs for existing nuclear plants are generally low, allowing many to remain competitive in electricity markets. Yet, economic pressures from cheap natural gas and subsidized renewables have led to the early closure of some nuclear plants in deregulated markets. This underscores the importance of market structures and energy policies in determining nuclear viability.

Government support is crucial in shaping nuclear economics. Loan guarantees, tax incentives, and other policy measures can significantly improve the financial outlook for nuclear projects. Many countries pursuing nuclear power provide some form of state backing or market support, though the level and nature of government involvement vary widely.

The long-term economics of nuclear power are also influenced by waste management costs, decommissioning expenses, and potential carbon pricing. While these factors create additional financial considerations, they may also enhance nuclear's relative position as climate policies evolve.

Emerging technologies like small modular reactors (SMRs) aim to address some of the economic challenges facing large-scale nuclear plants. Their modular design and potential for factory fabrication could reduce construction costs and timelines. However, SMRs are still in the early stages of development, and their economic performance remains to be proven at a commercial scale.

We conclude that the economic viability of nuclear power is complex and variable. While nuclear power faces significant cost challenges, particularly for new constructions in some markets, it can be economically competitive under certain conditions. Government policies, market structures, technological advancements, and climate considerations will continue to shape the economic prospects of nuclear power in the coming decades.

References

1. 5World Nuclear Association. Economics of Nuclear Power. (2023, Sept. 29). https://world-nuclear.org/information-library/economic-aspects/economics-of-nuclear-power
2. Marketplace. Adams, K. (2023, July 31). New Georgia reactor is a test case for nuclear power. https://www.marketplace.org/2023/07/31/new-georgia-reactor-nuclear-power/
3. Healthy Environment Alliance of Utah. (2019, May). Analyzing the cost of small modular reactors and alternative power portfolios. https://www.utah.gov/pmn/files/515877.pdf
4. Kutak Rock LLP. (2023, Nov. 8). What is the Cost of Carbon-Neutral Baseload 24/7 Power Sources? Federal Energy Experts Release a Comparative Paper for Utilities. https://www.kutakrock.com/newspublications/news/2023/november/paper-carbonneutral-baseload-power-sources

5. U.S. Energy Information Administration. (2022). Levelized costs of new generation resources in the Annual Energy Outlook 2022. In *U.S. Energy Information Administration*. https://www.eia.gov/outlooks/aeo/pdf/electricity_generation.pdf

6. Lazard. (2024, June). Lazard's Levelized Cost of Energy Analysis—Version 17.0. https://www.lazard.com/media/xemfey0k/lazards-lcoeplus-june-2024-_vf.pdf

7. Change Oracle. Matthews, R. (2022, July 20). Nuclear power versus renewable energy. https://changeoracle.com/2022/07/20/nuclear-power-versus-renewable-energy/

8. Higher renewables costs, uncertainty show need for diverse energy resources: Lazard report. (2024, June 28). *Utility Dive*. https://www.utilitydive.com/news/higher-renewable-energy-costs-lazard-lcoe-storage-hydrogen/720177/

Chapter 12

Nuclear Power and National Security: Global Perspectives

In this chapter, we explore the intricate relationship between nuclear power and national security, examining global perspectives on this critical issue. While nuclear power offers significant benefits in combating global warming, it also raises serious concerns related to proliferation, terrorism, and geopolitical stability. By analyzing the various strategies and viewpoints adopted by countries worldwide, we can better understand the complex interplay between nuclear power and national security.

What is a Nuclear Weapon?

A nuclear weapon is an explosive device that derives its immense destructive force from nuclear fission, nuclear fusion, or a combination of both. These weapons, often referred to as atom bombs, nuclear bombs, or simply nukes, fall into two broad categories: fission weapons and fusion-based designs, also known as thermonuclear weapons or hydrogen bombs.

Nuclear weapons unleash tremendous explosive force, measured in kilotons (1,000 tons of TNT) and megatons (1,000,000 tons of TNT), along with intense heat and radiation. They are the most devastating weapons on Earth, capable of causing unparalleled death, destruction, and long-term environmental damage.

The Nuclear Non-Proliferation Treaty (NPT)

The Nuclear Non-Proliferation Treaty remains the cornerstone of global efforts to prevent the spread of nuclear weapons while promoting the peaceful use of nuclear energy. Opened for signature in 1968 and entering into force in 1970, the NPT has been signed by 191 states, making it one of the most widely accepted arms control agreements. It establishes a framework for disarmament, non-proliferation, and the peaceful exchange of nuclear technology, playing a crucial role in maintaining global security and stability.

Designation of Nuclear-Weapon States (NWS)

The NPT designates five countries as nuclear-weapon states: the United States, Russia, China, France, and the United Kingdom. These NWS are also permanent members of the United Nations Security Council and play significant roles in shaping global nuclear policies. The designation of these five countries is based on their status as the only nations that had manufactured and detonated a nuclear explosive device before January 1, 1967, as stipulated by the NPT.

Current Nuclear Arsenals

- **United States:** Approximately 1,419 strategic deployed warheads, with a total active military stockpile of around 3,750 warheads. Including retired warheads awaiting dismantlement, the total is about 5,044 warheads.

- **Russia:** Approximately 1,549 strategic deployed warheads, with a military stockpile of about 4,380 nuclear warheads. Including retired warheads awaiting dismantlement, the total is around 5,500 warheads.

- **China:** Estimates suggest China possesses about 410-450 nuclear warheads.

- **France:** France currently has approximately 290-300 nuclear warheads in its military stockpile.

- **United Kingdom:** The UK maintains about 225 nuclear warheads.

Treaty Obligations and Cooperation

Under the NPT, nuclear-weapon states agree not to assist non-nuclear-weapon states (NNWS) in developing nuclear weapons, while NNWS commit to not pursuing them. All parties to the treaty agree to cooperate in the peaceful development of nuclear energy. The NPT promotes collaboration in the field of peaceful nuclear technology and ensures equal access to this technology for all States parties, with safeguards to prevent the diversion of fissile material for weapons use.

The NPT continues to be a pivotal instrument in global efforts to prevent nuclear proliferation, promote disarmament, and facilitate the peaceful use of nuclear energy. The designation of the five NWS and their current nuclear arsenals underscores the ongoing challenges and responsibilities in managing and reducing nuclear weapons globally.

Countries Outside the NPT

Several countries possess nuclear capabilities but are not recognized as nuclear-weapon states under the NPT. Israel, India, Pakistan, and North Korea have developed significant nuclear arsenals despite their non-recognition under the treaty. Their capabilities include:

- **Israel:** Estimated 90-400 warheads, maintaining a policy of deliberate ambiguity. Israel is believed to have diverse delivery systems, including aircraft, submarine-launched cruise missiles, and ballistic missiles.

- **India:** Approximately 160-164 warheads, with expanding delivery systems, including ballistic missiles and submarine-launched ballistic missiles. India is also developing Multiple Independently Targetable Re-entry Vehicle (MIRV) capability.

- **Pakistan:** Approximately 170 warheads, with an expanding nuclear arsenal and delivery systems, including ballistic and cruise missiles. Pakistan is also pursuing MIRV technology.

- **North Korea:** Estimates range from 30-50 warheads. North Korea is developing various delivery systems, including intercontinental ballistic missiles (ICBMs) and tactical nuclear warheads.

All four countries are actively modernizing their nuclear arsenals and delivery systems. Their status outside the NPT framework presents ongoing challenges for global non-proliferation efforts, including increased regional tensions, potential arms races, and complications in global disarmament efforts. These developments underscore the need for continued diplomatic efforts to address nuclear proliferation concerns and promote global disarmament.

Safeguards and Verification

The International Atomic Energy Agency (IAEA) plays a crucial role in implementing safeguards under the NPT to ensure that nuclear material is not diverted to weapons. These safeguards are embedded in legally binding agreements, primarily Comprehensive Safeguards Agreements and Additional Protocols.

Comprehensive Safeguards Agreements (CSAs): Under CSAs, the IAEA has the right and obligation to verify that all nuclear material in a non-nuclear-weapon state (NNWS) is used exclusively for peaceful purposes. As of May 2023, the IAEA has concluded CSAs with 182 states, allowing the agency to conduct various verification activities, including:

- **On-site Inspections:** Regular inspections to verify the correctness and completeness of a state's declared nuclear material and activities.
- **Environmental Sampling:** Collection of samples at facilities to detect undeclared nuclear material and activities.
- **Remote Monitoring:** Use of tamper-proof seals, cameras, and other surveillance technologies to monitor nuclear material movements.

Additional Protocols (APs): APs significantly enhance the IAEA's ability to detect undeclared nuclear material and activities. They include measures such as:

- **Expanded Access:** Greater access to information and locations within a state.
- **Unannounced Inspections:** Increased use of unannounced inspections to ensure compliance.

- **Enhanced Information Analysis:** Comprehensive evaluation of information from state declarations, IAEA verification activities, and open sources.

The five nuclear-weapon states under the NPT have concluded agreements with the IAEA, allowing it to apply safeguards to nuclear material in those states' facilities.

- **Item-Specific Safeguards:** For states not party to the NPT, such as India, Israel, and Pakistan, the IAEA implements item-specific safeguards agreements. These agreements ensure that nuclear material and facilities are used exclusively for peaceful purposes and not for the manufacture of nuclear weapons.

- **Coordination and Training:** The IAEA collaborates closely with state and regional systems for nuclear material accounting and control. It also provides enhanced training for IAEA inspectors and state personnel responsible for safeguards implementation.

The Role and Impact of IAEA Safeguards

The IAEA's safeguards system functions as a confidence-building measure, an early warning mechanism, and a trigger for international responses if necessary. By verifying that states adhere to their nonproliferation commitments, the IAEA plays an indispensable role in the global effort to prevent the spread of nuclear weapons and promote the peaceful use of nuclear energy.

The Strategic Arms Reduction Treaty (START)

The Strategic Arms Reduction Treaty series are pivotal bilateral agreements between the United States and the Soviet Union/Russia, aimed at reducing and limiting strategic offensive arms. The first of these, START I, was signed in 1991 by the U.S. and the Soviet Union and came into force in 1994, following the dissolution of the Soviet Union. This treaty required both nations to reduce their deployed strategic nuclear warheads to 6,000 and their delivery vehicles to 1,600. START I remained in effect until its expiration in 2009.

The most recent agreement, known as New START, was signed in 2010 and came into effect in 2011. This treaty limits each side to 1,550 deployed nuclear warheads and 700 deployed delivery systems. In 2021, New START was extended until February 2026. The START treaties have been instrumental in reducing nuclear arsenals from Cold War levels, incorporating verification measures such as on-site inspections to ensure compliance. These agreements have contributed significantly to strategic stability, providing a framework for predictability and reducing the risk of nuclear conflict. Overall, the START treaties represent significant milestones in nuclear arms control efforts between the two largest nuclear powers.

Impact of Russia's Invasion of Ukraine on the New START Treaty

Russia's invasion of Ukraine in 2022 has profoundly affected the New START Treaty, the last remaining nuclear arms control agreement between the United States and Russia. In response to the deteriorating relations, Russian President Vladimir Putin announced the suspension of Russia's participation in the treaty. While this suspension is not a complete withdrawal, it effectively halts the treaty's implementation, including the crucial on-site inspections and data exchanges that ensure compliance and transparency.

Russia justified its suspension on several grounds, including accusations that the U.S. failed to comply with certain treaty obligations. Russia also claimed that Western sanctions and support for Ukraine impeded Russian inspectors' ability to conduct their duties. Additionally, Russia has voiced broader geopolitical grievances, arguing that arms control cannot be separated from the current strategic and military realities, particularly the conflict in Ukraine and the perceived threat from NATO.

In retaliation, the United States has taken measures such as revoking visas for Russian nuclear inspectors and halting information sharing on missile statuses and test launches. These actions are framed as lawful countermeasures in response to what the U.S. describes as Russia's persistent violations of the treaty.

The suspension of the New START Treaty raises significant concerns about the future of nuclear arms control and the potential for a new arms race. The treaty, which limits each country to 1,550 deployed nuclear warheads and 700 deployed missiles and bombers, is set to expire in February 2026. With the suspension, the mechanisms for ensuring compliance and transparency are severely weakened, increasing the risk of misunderstandings and strategic instability. The invasion of Ukraine has not only strained U.S.-Russia relations but also jeopardized key frameworks for nuclear arms control, highlighting the broader implications of geopolitical conflicts on global security agreements.

Geopolitical Considerations

The pursuit of nuclear power has significant geopolitical implications, particularly for developing countries. Several factors influence these dynamics:

- **International Partnerships:** Many countries partner with established nuclear powers to develop their own nuclear energy programs. This can create long-term diplomatic and economic ties, as well as potential dependencies.

- **Regional Dynamics:** The development of nuclear capabilities can alter regional power balances and influence relationships between neighboring countries.

- **Competition Among Major Powers:** The United States, Russia, and China are engaged in a geopolitical competition for influence in the global nuclear energy market. This competition extends beyond commercial interests to issues of foreign policy and strategic influence.

- **Energy Security:** For many developing countries, nuclear power represents an opportunity to enhance energy security and reduce dependence on fossil fuels.

The international community closely monitors the development of nuclear programs to ensure they align with non-proliferation goals and regional stability.

Emerging Nuclear Power Programs

Several developing countries are exploring or actively pursuing nuclear power, presenting both opportunities and challenges:

- **Southeast Asian Countries:** Singapore, Thailand, the Philippines, Indonesia, and Malaysia have expressed interest in nuclear energy.

- **Central Asian, African, and Eastern European Countries:** Many nations in these regions are considering accelerated timelines for nuclear energy development.

These emerging nuclear powers must navigate complex technical, financial, and political considerations, including safety culture, transparency, and regional cooperation. Balancing their energy needs with non-proliferation obligations is crucial. International collaboration and assistance, along with stringent safeguards, are essential in ensuring the responsible development and deployment of nuclear energy.

Counterterrorism and Nuclear Security

Ensuring the security of nuclear facilities and materials against terrorist threats is a critical global concern. International efforts in this area focus on several key initiatives:

- Physical Protection Measures for Nuclear Facilities

- Enhanced Security for the Transportation of Nuclear Materials

- Information Sharing and Cooperation Among Countries

- Capacity Building and Training Programs for Nuclear Security Personnel

Safeguarding Nuclear Power Plants: Counterterrorism Measures

Counterterrorism measures are essential for safeguarding nuclear power plants due to the potentially catastrophic consequences of a successful terrorist attack. These measures encompass a comprehensive range of strategies designed to protect against various threats, including physical attacks, cyber-attacks, and insider threats. The IAEA plays a central role in coordinating these efforts and promoting best practices in nuclear security.

Here are the key counterterrorism measures implemented to safeguard nuclear power plants:

Physical Security Measures

- **Robust Structural Design:** Nuclear power plants are constructed with highly robust containment structures designed to withstand significant impacts, including those from aircraft crashes. These structures are among the strongest and most impenetrable defenses.

- **Multiple Security Barriers:** Facilities are protected by multiple layers of physical barriers, including fences, walls, and controlled access points, all designed to delay and deter unauthorized access.

- **Armed Security Personnel:** Highly trained and armed security forces are stationed at nuclear facilities. These personnel undergo rigorous training and participate in regular force-on-force exercises to ensure they are prepared for potential attacks.

- **Increased Standoff Distances:** To prevent vehicle-borne improvised explosive device attacks, increased standoff distances are maintained around critical areas of the plant.

Technological and Cybersecurity Measures

- **Surveillance and Monitoring:** Advanced surveillance systems, including closed-circuit television (CCTV) and motion detectors, are employed to continuously monitor the perimeter and sensitive areas within the plant.

- **Cybersecurity Protocols:** With the growing threat of cyber-attacks, nuclear plants have implemented stringent cybersecurity measures to protect their control systems and sensitive information. This includes firewalls, intrusion detection systems, and regular security audits.

Operational and Procedural Measures

- **Background Checks and Access Control:** Comprehensive background checks are conducted for all personnel, including temporary workers and visitors. Access to sensitive areas is strictly controlled and monitored.

- **Emergency Response Plans:** Detailed and well-coordinated emergency response plans are in place to manage potential incidents. These plans include evacuation procedures, medical assessment and treatment, and decontamination capabilities.

- **Regular Drills and Exercises:** Nuclear facilities conduct regular security drills and exercises, including simulations of terrorist attacks, to ensure preparedness and identify areas for improvement.

International and National Cooperation

In the United States, the Nuclear Regulatory Commission (NRC) works closely with federal agencies such as the Department of Homeland Security (DHS), the Federal Aviation Administration (FAA), and the Department of Defense (DOD) to ensure the protection of nuclear facilities. This collaboration is critical in developing and implementing comprehensive security measures that address a wide range of potential threats.

Internationally, the International Atomic Energy Agency (IAEA) provides essential guidance through its Nuclear Security Series, establishing international consensus on nuclear security practices. This guidance not only helps states implement effective security measures but also fosters cooperation across borders, ensuring a unified global approach to nuclear security.

Effective information sharing between nuclear facilities and relevant authorities is vital for timely threat assessment and response. This includes the dissemination of intelligence on potential threats and the exchange of best practices, which enhances the overall security framework.

Mitigation Strategies for Nuclear Security

To safeguard nuclear facilities, several critical mitigation strategies are implemented:

Redundant Safety Systems: Nuclear plants are equipped with redundant safety systems designed to protect the reactor core and containment structures, ensuring their integrity even in the event of an attack.

Dry-Cask Storage: To mitigate the risks associated with spent fuel pools, additional dry-cask storage capacity is employed, as it is less vulnerable to external attacks.

Underground Facilities: Future reactors and spent fuel storage facilities are being considered for underground placement, further enhancing their protection against potential threats.

Enhanced Security Measures Post-9/11

On September 11, 2001, the world witnessed an unprecedented act of terror when 19 extremists from the al Qaeda terrorist group hijacked four commercial airplanes. Two of these planes were deliberately flown into the North and South Towers of the World Trade Center in New York City, leading to the catastrophic collapse of both towers. A third plane struck the Pentagon in Arlington, Virginia, causing significant damage to the heart of the U.S. defense establishment. The fourth plane, United Airlines Flight 93, was headed for another target, but brave passengers fought back against the hijackers, resulting in the plane crashing into a field in Pennsylvania.

Nearly 3,000 people from 93 countries lost their lives in these attacks, marking the deadliest terrorist act on American soil. The 9/11 attacks not only shattered lives but also profoundly impacted global security policies and the way nations approach counterterrorism.

The terrorist attacks of 9/11 served as a wake-up call, leading to significant security enhancements at nuclear plants:

- **Increased Armed Security Personnel:** More highly trained and qualified security forces are now stationed at nuclear facilities.

- **Stricter Access Controls:** Enhanced background checks and more rigorous access control measures have been implemented.

- **Improved Physical Barriers:** Multiple layers of fencing and controlled access points provide greater physical security.

- **Expanded Standoff Distances:** Increased standoff distances help protect against vehicle-borne threats.

- **Advanced Surveillance Systems:** Improved CCTV, motion detectors, and other surveillance technologies enhance monitoring capabilities.

- **Enhanced Cybersecurity Protocols:** Stricter cybersecurity measures protect control systems from potential cyber-attacks.

- **Early Warning Systems:** Coordination with the FAA and NORAD ensures early detection and response to airborne threats.

- **Updated Design Basis Threat (DBT):** The DBT has been revised to reflect realistic terrorist scenarios, ensuring preparedness for a range of threats.

- **Enhanced Force-on-Force Drills:** Regular force-on-force security drills are conducted to maintain a high level of readiness.

- **Improved Emergency Response Plans:** Enhanced coordination with local authorities ensures effective response to any incidents.

- **Stricter Fitness-for-Duty Requirements:** Security personnel are held to higher fitness standards to ensure their capability to respond to threats.

- **Increased Interagency Cooperation:** Federal agencies now work more closely together on threat assessment and response strategies.

- **Comprehensive Vulnerability Assessments:** Regular assessments are conducted to identify and address potential vulnerabilities.

These measures reflect a layered and comprehensive approach to nuclear plant security, addressing physical, cyber, and operational aspects. The NRC continuously evaluates and updates these requirements to adapt to evolving threats and emerging technologies. By implementing these robust counterterrorism measures, nuclear power plants significantly reduce the risk of terrorist attacks, ensuring the safety and security of the public and the environment.

Conclusion: The Intersection of Nuclear Power and National Security

Nuclear power and national security are inextricably linked, demanding a careful balance between harnessing the immense benefits of nuclear energy and addressing the serious risks of proliferation and terrorism. The Nuclear Non-Proliferation Treaty (NPT), alongside safeguards, verification measures, counterterrorism efforts, and geopolitical considerations, profoundly shapes the global perspective on nuclear power and security. By fostering international cooperation and adhering to non-proliferation commitments, nations can leverage the potential of nuclear power while striving toward a safer and more secure world.

As the intersection of nuclear power and national security continues to evolve within a complex global landscape, it becomes increasingly clear that while nuclear energy offers significant advantages in combating climate change and bolstering energy security, it also presents critical challenges related to proliferation, security, and geopolitical stability. International cooperation, robust safeguards, and a steadfast commitment to non-proliferation principles are essential to harnessing the potential of nuclear power while mitigating these associated risks.

The enhanced security measures implemented post-9/11 have markedly improved the safety and security of nuclear facilities, incorporating increased armed security personnel, advanced surveillance systems, and stringent cybersecurity protocols. These comprehensive counterterrorism strategies have significantly reduced the risk of terrorist attacks, ensuring the safety of the public and the environment.

The Strategic Arms Reduction Treaty (START) and its successors have been pivotal in reducing nuclear arsenals and promoting stability between the United States and Russia. These agreements established critical frameworks for verification, inspections, and data exchanges, enhancing transparency in strategic nuclear forces.

However, Russia's invasion of Ukraine in 2022 led to a sharp deterioration in U.S.-Russia relations, culminating in Russia's suspension of participation in the New START Treaty in February 2023. This suspension halted vital on-site inspections and data exchanges, prompting U.S. countermeasures and raising significant concerns about the future of nuclear arms control. The increased risk of misunderstandings and strategic instability underscores the fragility of such agreements in the face of escalating geopolitical tensions.

With New START set to expire in February 2026 and its compliance mechanisms weakened, uncertainty looms over the future of arms control. This development highlights the urgent need for renewed diplomatic efforts to preserve and strengthen arms control frameworks in an increasingly complex international environment.

As the global community navigates these challenges, it must balance the pursuit of clean energy with the imperative of maintaining international security and stability. The future of nuclear power will depend on the ability of nations to cooperate effectively, address shared concerns, and uphold the principles of the non-proliferation regime. Through international collaboration, adherence to stringent security measures, and a continued commitment to arms control agreements, it is possible to harness the benefits of nuclear energy while ensuring a secure and stable global environment. Yet, the recent setbacks in arms control underscore the need for renewed diplomatic efforts and innovative approaches to address the evolving challenges in the nuclear arena.

References:

1. United Nations Office for Disarmament Affairs. (n.d.). *Treaty on the Non-Proliferation of Nuclear Weapons (NPT)*. https://disarmament.unoda.org/wmd/nuclear/npt/
2. Reaching Critical Will. (2024). *2024 NPT Preparatory Committee.* https://reachingcriticalwill.org/disarmament-fora/npt/2024

3. International Atomic Energy Agency (IAEA). (n.d.). *IAEA Safeguards Overview.* https://www.iaea.org/publications/fact-sheets/iaea-safeguards-overview

4. World Population Review. (2024). *Nuclear Weapons by country 2024.* https://worldpopulationreview.com/country-rankings/nuclear-weapons-by-country

5. Arms Control Association. (April 2023). *Understanding the dispute over new START.* https://www.armscontrol.org/act/2023-04/news/understanding-dispute-over-new-start

6. Nuclear Regulatory Commission. (March 12, 2020). *Frequently asked questions about security assessments at nuclear power plants.* https://www.nrc.gov/security/faq-security-assess-nuc-pwr-plants.html

Chapter 13

Public Perception and Education: Overcoming the Stigma of Nuclear Power

Now is the time to explore the critical role of public perception, education, and community engagement in shaping the future of nuclear power. Despite its potential as a green energy solution, nuclear power often faces public skepticism and stigma due to concerns about safety, security, waste management, and past accidents. By addressing these concerns, including the NIMBY (Not In My Backyard) phenomenon, and promoting accurate information, we can foster a more informed and supportive public opinion toward nuclear power in the fight against global warming.

Understanding Public Perception

Public perception of nuclear power is influenced by a range of factors, including historical events, media portrayal, personal beliefs, and concerns about proximity to nuclear facilities. Incidents like the Chernobyl and Fukushima accidents have profoundly shaped public opinion, contributing to the lingering stigma around nuclear energy. Additionally, NIMBY concerns—where communities resist having nuclear facilities or waste disposal sites anywhere near their homes—further complicate public acceptance. To effectively address these concerns and misconceptions, it is essential to understand the roots of public perception.

Lessons Learned from Nuclear Plant Accidents

Historical events such as the Chernobyl and Fukushima nuclear accidents have cast long shadows over public opinion. However, it is crucial to compare the actual adverse human effects of these nuclear accidents with the ongoing and widely accepted adverse effects of fossil fuel energy production.

Three Mile Island (1979)

- **Immediate Impact:** Widespread public fear and uncertainty led to approximately 140,000 people, primarily pregnant women and young children, being temporarily evacuated.

- **Long-term Health effects:** Studies conducted by the National Cancer Institute (NCI) in the years following the accident found no evidence of long-term health impacts or increased cancer rates among the general population.

Chernobyl (1986)

- **Immediate Impact:** The Chernobyl disaster resulted in 31 immediate deaths among workers and first responders due to acute radiation syndrome.

- **Long-term Health Effects:** Over 6,000 cases of thyroid cancer have been reported among those exposed as children or adolescents at the time of the accident, with more cases expected.

- **Wider Health Consequences:** The accident led to increased risks of leukemia, cardiovascular diseases, and cataracts among cleanup workers. Psychological effects, including high anxiety and PTSD, were significant among the affected populations.

Fukushima (2011)

- **Immediate Impact:** The Fukushima disaster caused no direct radiation-related deaths, although a lung cancer death 4 years later in a plant worker was attributed to radiation exposure.

- **Health Assessments:** According to the United Nations Scientific Committee on the Effects of Atomic Radiation (UNSCEAR), no observable negative health effects for the public are expected due to radiation exposure. However, the evacuation and relocation efforts resulted in increased mortality among the elderly, higher rates of non-communicable diseases, and significant psychological distress, including PTSD and depression.

Fossil Fuels: The Ongoing Adverse Effects

- **Annual Death Toll:** Air pollution from burning fossil fuels is responsible for more than 8 million deaths annually worldwide. This is vastly higher than the combined one-time direct deaths from Chernobyl and Fukushima.

- **Health Problems:** Fossil fuel pollution causes a wide range of health issues, including respiratory infections, heart disease, cancer, and asthma. These impacts are severe, ongoing, and contribute to a major public health crisis.

A Comparative Perspective

To put this into perspective, in over 18,500 cumulative reactor-years of commercial nuclear power operation across 36 countries, only three major nuclear accidents have occurred: Three Mile Island, Chernobyl, and Fukushima. Six decades of nuclear energy use demonstrate that it is one of the safest methods of generating electricity. The risk of accidents at nuclear power plants is low and continues to decrease with advances in reactor design. Furthermore, the consequences of a nuclear accident or even a terrorist attack are minimal compared to the risks we regularly accept from other energy sources.

By contrast, fossil fuels are responsible for millions of deaths each year due to air pollution—a staggering toll that far surpasses the fatalities from all nuclear incidents combined. When we also consider the devastating health effects of global warming, the number of deaths, injuries, and diseases linked to fossil fuel use skyrockets. This underscores the severe and ongoing public health crisis fueled by reliance on fossil energy.

Educating the public about the relative safety of nuclear power—especially in comparison to the grave health risks posed by fossil fuels—can play a pivotal role in shifting perceptions. Nuclear energy must be seen as a critical part of a sustainable, low-carbon future that prioritizes human health and environmental preservation.

NIMBY Concerns in Nuclear Waste Management and Nuclear Plant Siting

Addressing NIMBY (Not In My Backyard) concerns is a crucial aspect of both managing the disposal of radioactive waste and the siting of nuclear power plants. NIMBY refers to the opposition or reluctance of communities to have nuclear facilities—whether for waste disposal or energy production—located near their homes or in their immediate vicinity. These concerns are valid and often stem from fears about potential risks, including safety, environmental impact, and the perceived stigma associated with nuclear energy. Addressing these concerns requires a multi-faceted approach that prioritizes open communication, transparency, and active community engagement.

Strategies to Mitigate NIMBY Opposition

- **Early, Open, and Continuous Engagement**: Engage with local communities early in the planning process before decisions are made. Continuous dialogue and information sharing throughout the project lifecycle can help build trust and address concerns proactively, whether the project involves nuclear waste disposal or the construction of a new nuclear power plant.

- **Clear and Accessible Information:** Provide clear, easy-to-understand, and accurate information about the project, its potential impacts, and the safety measures being implemented. Help neighbors grasp the rigorous processes involved in the selection and evaluation of disposal sites and plant locations, including geological, environmental, hydrological, and safety assessments.

- **Address Misconceptions:** Directly address common misconceptions and fears surrounding nuclear waste storage and nuclear power plant operations. Provide factual information from credible sources to counter misinformation and alleviate concerns.

- **Inclusive Decision-Making:** Involve local communities in the decision-making process by soliciting their input, concerns, and suggestions. Demonstrate that their voices are heard and valued, especially when considering the siting of nuclear facilities.

- **Community Advisory Boards:** Establish community advisory boards or committees that include diverse stakeholders, such as local residents, elected officials, environmental groups, and subject matter experts. These boards can serve as platforms for open dialogue, information sharing, and collaborative problem-solving.

- **Local Benefits and Compensation:** Explore ways to provide tangible benefits to host communities, such as economic incentives, infrastructure improvements, or investments in local projects. Emphasize the high-paying jobs that will be created and the low-carbon and potentially lower cost electricity that will be available in the area. Fair compensation can help address concerns about potential negative impacts and can be particularly persuasive when communities are considering whether to accept a nearby nuclear plant or waste facility.

Building Trust and Credibility

Building trust and credibility with local communities is essential for addressing NIMBY concerns. Involving independent, third-party experts and regulatory bodies to oversee the project can help build credibility and alleviate concerns about potential conflicts of interest. Maintaining transparency throughout the project, regularly sharing updates, reports, and monitoring data with the public, and establishing clear accountability measures are crucial steps in this process.

Addressing Concerns About Nuclear Power Plants

In addition to concerns about waste disposal, communities may also have specific worries about the construction of nuclear power plants near their homes. These concerns typically center around the potential for accidents, environmental impact, and the long-term safety of living near a nuclear facility.

- **Safety and Accident Risk:** It is crucial to communicate the robust safety measures and advanced technologies that modern nuclear plants employ to prevent accidents. Discussing the lessons learned from past incidents and how they have led to the development of safer, more resilient designs can help alleviate fears.

- **Environmental Impact:** Address concerns about the potential environmental impact of a nuclear power plant, including its effects on local ecosystems and water resources. Providing evidence of the minimal environmental footprint of modern nuclear plants compared to fossil fuel alternatives can help reshape public opinion.

- **Long-Term Safety:** Ensuring the public that nuclear facilities are subject to stringent regulatory oversight and that long-term safety is a top priority is essential. This includes discussing emergency preparedness plans, the training of plant personnel, and the role of continuous monitoring and inspections.

The Role of Media in Shaping Public Perception

The media plays a powerful role in shaping public perception of nuclear power, often focusing on dramatic events rather than the everyday safety and efficiency of nuclear energy production. By engaging with the media and ensuring balanced reporting, we can provide accurate, science-based information to the public. This effort is crucial in overcoming public skepticism and fostering a more informed perspective on the benefits of nuclear energy.

Government and Industry Transparency

Transparency from both the government and the nuclear industry is essential in building public trust. Openly sharing information about nuclear operations, safety protocols, and incident responses demonstrates a commitment to public safety and environmental stewardship. This transparency is vital for dispelling myths, addressing public concerns, and fostering a positive perception of nuclear power.

Case Studies of Successful Public Engagement

Examining case studies of countries or regions where public engagement has successfully shifted perception in favor of nuclear power can provide valuable insights. For example, France's long-standing public acceptance of nuclear power, with its lower electricity costs and less air pollution, and Finland's progress in waste management and public trust serve as models for other countries. These examples illustrate how effective public engagement can lead to widespread support for nuclear energy.

The Influence of Social Media and Digital Platforms

In today's digital age, social media and online platforms play a significant role in shaping public opinion. Developing strategies to effectively use these platforms for educating the public, combating misinformation, and engaging younger generations is essential. Creating accessible content—such as infographics, videos, and interactive tools—can reach a wider audience and foster a deeper understanding of nuclear power.

Safety and Risk Communication

Safety is a primary concern when it comes to modern nuclear power. It is vital that we clearly communicate the rigorous safety measures and regulatory frameworks in place to protect both people and the environment. Open and transparent communication about safety practices, emergency preparedness, and risk assessment can help build trust and alleviate public concerns.

Waste Management and Disposal

Waste management and disposal are understandably significant public concerns. It is crucial to educate the public about the strategies and advancements in waste management, including storage, reprocessing, and deep geological repositories. Providing accurate information about the long-term safety of waste disposal can help dispel misconceptions and ease public anxieties.

Public Engagement and Dialogue

Engaging the public in meaningful dialogue is essential for addressing concerns and dispelling myths about nuclear power. Public forums, educational campaigns, and community involvement provide opportunities for open discussions, allowing for the exchange of information, perspectives, and concerns. Engaging with stakeholders and actively listening to their views can help build trust and understanding.

The Role of Education

Public education is vital in dispelling myths and promoting accurate information about nuclear power. Incorporating nuclear energy topics into school curricula, organizing educational programs, and promoting scientific literacy can help foster a more informed and supportive public opinion. Encouraging critical thinking and providing access to reliable resources can empower individuals to make informed decisions about nuclear energy.

Long-Term Benefits of Nuclear Power

Beyond addressing concerns, it is essential to emphasize the long-term benefits of nuclear power as a sustainable energy source. Nuclear energy has the potential to complement renewable sources, providing reliable, low-carbon power that is crucial for achieving global climate goals. Framing nuclear power as an indispensable component of a future low-carbon energy mix will further strengthen its case for broader acceptance.

Policy Recommendations for Public Engagement

To effectively shift public perception, specific policy recommendations should be considered. Governments should increase funding for public education campaigns, provide incentives for research and development in nuclear safety, and foster international cooperation on best practices in nuclear energy communication. These steps are essential for creating an informed and supportive public that understands the vital role nuclear power plays in our sustainable energy future.

Empowering Change: How You Can Influence Nuclear Energy Policy for a Sustainable Future

In the fight against global warming, individual action is often perceived as limited to personal lifestyle changes, such as reducing energy consumption or supporting renewable energy initiatives. However, when it comes to shaping the future of nuclear energy policy, individuals have a unique and powerful role to play. By engaging in informed advocacy, participating in public discourse, and holding elected officials accountable, individuals can significantly influence the direction of nuclear energy policy, ensuring it supports a sustainable and low-carbon future. Here's how you can take meaningful action:

1. Educate Yourself and Others

The first step in influencing public policy is to be well-informed yourself. Understanding the scientific, economic, and safety aspects of nuclear energy allows you to engage in discussions and debates with confidence. Begin by reading credible sources, including government reports, academic publications, and expert opinions on nuclear energy. Many good references on these subjects are included after each chapter in this book. Attend public lectures, webinars, and community forums where nuclear energy is discussed.

Once you are informed, share your knowledge with others. Use social media, blogs, or community meetings to educate your peers about the benefits and challenges of nuclear energy. By spreading factual information, you can help counteract misinformation and build a base of support for nuclear energy as a key component of our sustainable energy future.

2. Engage with Your Elected Representatives

Your elected officials play a crucial role in shaping nuclear energy policy, and they need to hear from their constituents to understand public opinion. Start by identifying your local, state, and federal representatives who have influence over energy policy. Write letters, send emails, or make phone calls to express your support for nuclear energy and its role in combating climate change.

Be specific in your communication. Highlight recent advancements in nuclear technology, such as small modular reactors (SMRs) or improved waste management solutions and explain how these innovations can enhance safety and efficiency. Emphasize the economic benefits, including job creation and energy security, that nuclear energy can bring to your community and the nation.

You can also request meetings with your representatives to discuss nuclear energy policy in person. Bringing together a group of like-minded individuals or forming a local advocacy group can amplify your voice and demonstrate to your representatives that there is strong public support for nuclear energy.

3. Participate in Public Comment Periods and Hearings

When new nuclear projects or policies are proposed, regulatory agencies often hold public comment periods or hearings to gather input from citizens. These are critical opportunities for you to influence the decision-making process. Stay informed about upcoming comment periods by following the websites of agencies such as the Nuclear Regulatory Commission (NRC) or state-level environmental and energy departments.

Prepare thoughtful, evidence-based comments that support nuclear energy initiatives while addressing potential concerns such as safety or environmental impact. If possible, attend public hearings in person or virtually to voice your support. Your participation not only contributes to the policy-making process but also signals to decision-makers that there is public backing for nuclear energy.

4. Support Pro-Nuclear Advocacy Organizations

Numerous organizations advocate for nuclear energy at the local, national, and international levels. By joining or supporting these groups, you can help amplify their efforts to influence public policy. These organizations often have the resources and expertise to engage with policymakers, run public awareness campaigns, and challenge anti-nuclear legislation.

Consider donating to or volunteering with organizations that align with your views on nuclear energy. These groups often need support for lobbying efforts, educational outreach, and grassroots organizing. By contributing your time, money, or skills, you can help these organizations build a stronger case for nuclear energy in policy discussions.

5. Vote for Candidates Who Support Nuclear Energy

Your vote is one of the most direct ways to influence nuclear energy policy. Research candidates' positions on energy policy, particularly their stance on nuclear energy, before elections. Support and vote for candidates who recognize the importance of nuclear energy in stopping global warming by achieving a low-carbon, sustainable future and who are committed to advancing its development.

Beyond voting, you can also participate in campaign activities such as canvassing, phone banking, or hosting events for pro-nuclear candidates. Helping to elect officials who prioritize nuclear energy can lead to significant policy shifts at both the local and national levels.

6. Foster Community Dialogue and Support

Building local support for nuclear energy is essential for influencing broader policy decisions. Start by engaging with your community through town halls, school boards, or neighborhood associations. Organize discussions or forums on nuclear energy, inviting experts to present and answer questions.

By fostering an open dialogue within your community, you can address common concerns and misconceptions about nuclear energy. Highlight the environmental and economic benefits that a local nuclear project could bring, and work to build a coalition of supporters who can advocate for nuclear energy in local decision-making processes.

7. Use Social Media to Advocate for Nuclear Policy

Social media platforms are powerful tools for raising awareness and mobilizing support for nuclear energy. Use your social media accounts to share information, articles, and personal insights about the benefits of nuclear energy. Engage with influencers, policymakers, and advocacy groups to broaden the reach of your message.

Consider starting or joining online communities that focus on nuclear energy advocacy. By participating in discussions, sharing content, and encouraging others to take action, you can help create a groundswell of public support that influences policymakers and the media.

8. Advocate for Educational Programs and Curricula

Supporting the inclusion of nuclear energy topics in educational curricula is a long-term strategy for influencing public policy. Advocate for schools, universities, and technical programs to offer courses and degrees in nuclear science and engineering. Encourage the development of educational resources that provide balanced, factual information about nuclear energy.

By fostering a well-educated generation of future leaders, you contribute to a broader societal understanding of nuclear energy's role in achieving sustainability. This, in turn, can lead to more informed policy decisions as these individuals enter the workforce and engage in public discourse.

Making a Difference Through Advocacy

Influencing nuclear energy policy requires sustained effort, but as an informed and engaged citizen, your actions can have a significant impact. By educating yourself and others, engaging with representatives, participating in public processes, and supporting pro-nuclear candidates and organizations, you can help shape a sustainable energy future that includes nuclear power as a critical component. Your advocacy not only contributes to policy changes but also fosters a broader acceptance of nuclear energy as a viable and essential solution to global warming.

Successful Nuclear Power Adoption Across the Globe

As the world grapples with the escalating impacts of climate change, nuclear power emerges as a beacon of hope—offering a powerful solution to reduce carbon emissions and secure energy for the future. Several countries have successfully adopted nuclear energy, providing compelling case studies that illustrate its potential as a sustainable solution.

France: A Pioneer in Nuclear Energy

France is often cited as a leading example of successful nuclear power adoption. In response to the 1970s oil crisis, France made a strategic and bold decision to invest heavily in nuclear power, significantly reducing its dependence on fossil fuels. Today, over 70% of France's electricity is generated from nuclear energy, making it one of the world's largest producers of nuclear power. This transition has provided France with a stable and reliable energy source, positioning the country as a global leader in low-carbon energy production.

France's commitment to nuclear energy has dramatically reduced its greenhouse gas emissions, showcasing the environmental benefits of this energy source. Moreover, this shift has resulted in France having some of the lowest electricity prices in Europe, with household costs approximately 40% lower than the European average. To overcome public objections, France implemented a comprehensive public information campaign, emphasizing the economic benefits and energy independence that nuclear power would bring.

South Korea: Rapid Expansion and Technological Innovation

South Korea's journey into nuclear power began in the late 1970s, and since then, it has rapidly expanded its nuclear capacity. Today, the country operates 24 nuclear reactors, which generate about 30% of its electricity. South Korea's success is rooted in its focus on technological innovation and safety. The development of the APR1400, an advanced nuclear reactor design, highlights South Korea's commitment to improving nuclear technology. The country has not only achieved energy security but has also become a major exporter of nuclear technology, contributing to the global advancement of safe and efficient nuclear power.

South Korea's nuclear program has helped keep electricity prices relatively low, with industrial rates about 20-30% lower than in other developed countries. To address public concerns, the government has prioritized transparency, regular safety inspections, and community engagement programs, ensuring that the public is informed and involved.

Sweden: Balancing Nuclear Power with Renewable Energy

Sweden presents a unique case where nuclear power and renewable energy coexist in a balanced energy mix. Facing environmental challenges in the 1970s, Sweden embarked on an ambitious nuclear program. Today, nuclear energy accounts for approximately 40% of the country's electricity production. What sets Sweden apart is its successful integration of nuclear power with renewable sources such as hydropower and wind energy. This approach has enabled Sweden to maintain a low-carbon energy system while ensuring energy reliability. The Swedish model exemplifies how nuclear power can complement renewable energy, providing a stable backbone to a sustainable energy infrastructure.

Sweden's electricity prices are among the lowest in Europe, partly due to its nuclear and hydropower mix. Public acceptance of nuclear power in Sweden has been achieved through open dialogue, strict safety regulations, and a clear focus on the environmental benefits of nuclear energy.

United Arab Emirates: A Newcomer's Success

The United Arab Emirates (UAE) is a recent entrant into the nuclear power arena but has quickly established itself as a model for new nuclear power programs. In 2020, the UAE became the first Arab country to operate a nuclear power plant with the launch of the Barakah Nuclear Energy Plant. The plant is expected to provide up to 25% of the UAE's electricity once fully operational, significantly reducing the country's carbon footprint. The UAE's success is attributed to its strategic partnerships, robust regulatory framework, and commitment to safety. The Barakah project demonstrates that with careful planning and international collaboration, even countries new to nuclear power can achieve significant milestones.

While it's too early to fully assess the impact on consumer costs, the UAE expects nuclear power to help stabilize electricity prices in the long term. The UAE has proactively addressed public concerns through extensive community outreach, education programs, and by emphasizing the economic and environmental benefits of nuclear power.

China: Rapid Expansion and Technological Advancement

China has emerged as a formidable player in the global nuclear energy landscape, showcasing remarkable growth and innovation. Since the early 2000s, China has rapidly expanded its nuclear power capacity, with nuclear generation reaching 417.8 billion kilowatt-hours by 2022, accounting for 4.7% of total power generation. As of 2024, China boasts the world's largest capacity of nuclear reactors under construction, with projects totaling almost 30 gigawatts. The Chinese government has prioritized nuclear power as a key component of its clean energy strategy, contributing to the reduction of coal's share in power generation. China's success in nuclear adoption is characterized by its focus on domestic reactor design and construction, exemplified by the development of the Hualong One reactor.

To address public concerns, China has implemented strict safety measures and conducted extensive public education campaigns. While the impact on consumer electricity costs is still evolving, the massive scale of China's nuclear program is expected to contribute to long-term price stability and energy security.

Russia: Pioneer in Nuclear Technology and Global Exporter

Russia's nuclear power success story dates back to 1954 when it launched the world's first nuclear power plant for commercial purposes in Obninsk. Today, Russia operates 36 nuclear reactors with a total capacity of 26,802 MWe, demonstrating its long-standing commitment to nuclear energy. What sets Russia apart is its world-leading position in fast neutron reactor technology and its efforts to close the nuclear fuel cycle. Russia has also made nuclear exports a major policy and economic objective, with over 20 nuclear power reactors confirmed or planned for export construction. This success in both domestic implementation and international exports showcases Russia's expertise in nuclear technology.

Russia's nuclear program has contributed to relatively low electricity prices, with nuclear power often being one of the most cost-effective sources of electricity. To maintain public support, Russia has emphasized the economic benefits of nuclear power and its critical role in national energy security. However, the country continues to face challenges in retiring older units and balancing economic constraints with expansion plans.

Learning from Success

These real-world examples highlight the diverse pathways nations can take in adopting nuclear power. France, South Korea, Sweden, the UAE, China, and Russia each offer valuable lessons in strategic planning, technological innovation, and the integration of nuclear power with other energy sources. As we confront the global challenge of climate change, these success stories provide a roadmap for other nations considering nuclear energy as a key component of their sustainable energy strategies. By learning from these examples, we can better understand how nuclear power can play a crucial role in reducing global carbon emissions and securing a sustainable future for our planet.

Conclusion: Driving Change—Public Engagement, Education, and Policy for a Sustainable Nuclear Future

As we advance in the fight against global warming, the importance of nuclear power in our energy landscape cannot be overstated. The success stories of France, South Korea, Sweden, the UAE, China, and Russia demonstrate nuclear energy's powerful impact on reducing carbon emissions and ensuring energy independence. However, the widespread adoption of nuclear as a sustainable solution depends not just on technological progress, but also on public perception, education, and active community engagement.

To foster public support, we must address long-standing concerns about nuclear plant safety, waste management, and past accidents, as well as the common "Not In My Backyard" (NIMBY) mindset. Public opinion is shaped by a complex mix of historical events, media portrayals, personal beliefs, and proximity to nuclear facilities. Shifting this perception requires open, transparent communication that highlights the rigorous safety protocols, advances in waste management, and the relative safety of nuclear power compared to fossil fuels. We must also recognize the powerful role of media and digital platforms in shaping public sentiment and use these channels to spread accurate, science-based information.

Building trust through transparency is essential. Both government and industry must engage in continuous, honest dialogue with the public to dispel myths and foster confidence. Countries like France and South Korea have shown that, with the right approach, public acceptance of nuclear power is possible, even in the face of past stigmas. Addressing NIMBY concerns early, providing clear information, and involving local communities in decision-making are key steps toward building long-term support.

Public education is equally critical. By integrating nuclear energy into school curricula and promoting scientific literacy, we can equip future generations with the knowledge to make informed decisions about nuclear's role in mitigating climate change. This educational effort must be backed by policies that promote research, innovation, and international cooperation in nuclear safety and communication.

The earlier section, *Empowering Change: How You Can Influence Nuclear Energy Policy for a Sustainable Future*, emphasizes the role individuals can play in shaping nuclear policy. By becoming informed advocates, engaging with policymakers, and participating in public discourse, you can help drive decisions that support nuclear energy as part of a clean, reliable, and sustainable energy mix. Your involvement can significantly contribute to the broader acceptance of nuclear power as an essential solution to the climate crisis.

Nuclear power offers long-term benefits as a reliable, baseload, low-carbon energy source that can complement renewables in achieving global climate goals and halting our reckless march toward climate disaster. Framing nuclear power as indispensable to a low-carbon energy future strengthens its case for broader acceptance.

Achieving this shift in public perception requires a comprehensive strategy that includes transparent communication, active public engagement, addressing NIMBY concerns, robust educational initiatives, and individual participation. By aligning public understanding with the proven safety and potential of nuclear power, demonstrated by successes around the world, we can secure its place as a cornerstone of global efforts to combat climate change and ensure a clean energy future for generations to come.

References

1. United Nations Scientific Committee on the Effects of Atomic Radiation. (2018). *EVALUATION OF DATA ON THYROID CANCER IN REGIONS AFFECTED BY THE CHERNOBYL ACCIDENT.* https://www.unscear.org/unscear/uploads/documents/publications/Chernobyl_WP_2017.pdf

2. *How Chernobyl Jump-Started the global nuclear safety Regime.* (2019, September 12). U.S. GAO. https://www.gao.gov/blog/2019/09/12/how-chernobyl-jump-started-the-global-nuclear-safety-regime

3. *Safety of nuclear power reactors - World Nuclear Association.* (2022, March 2.). https://world-nuclear.org/information-library/safety-and-security/safety-of-plants/safety-of-nuclear-power-reactors

4. *Why some nations choose nuclear power - Kleinman Center for Energy Policy.* (2024, July 25). Kleinman Center for Energy Policy. https://kleinmanenergy.upenn.edu/research/publications/why-some-nations-choose-nuclear-power/

5. *Global survey finds high public support for nuclear: Nuclear Policies - World Nuclear News.* (2024, January 19). https://world-nuclear-news.org/Articles/Global-survey-finds-high-public-support-for-nuclea

Chapter 14

Towards a Greener Future: Nuclear Power is Essential to Save Humanity and Our Planet

One more subject needs to be mentioned as we close out this book, the rise of Artificial Intelligence (AI), particularly generative AI and large language models, is driving a huge surge in electricity demand. This increase has profound implications for electricity consumption, environmental sustainability, and the infrastructure required to support these technologies.

AI's Insatiable Demand for Electricity

AI systems, especially large language models like GPT-3 and GPT-4, and the more recent GPT-4o demand substantial computational power, which translates to high electricity consumption. For instance, training GPT-3 is estimated to use as much electricity as the annual consumption of about 130 U.S. homes. Even individual AI queries are energy-intensive; a single ChatGPT query consumes about 2.9 watt-hours of electricity, compared to just 0.3 watt-hours for a traditional Google search.

The International Energy Agency (IEA) projects that data center electricity consumption could double by 2026. Currently, data centers account for about 1-2% of global electricity use, but this could rise to

3-4% by the end of the decade. In the United States, projections suggest that data centers could consume up to 9% of the nation's electricity generation by 2030. To support this demand, U.S. utilities may need to invest around $50 billion in new generation capacity specifically for data centers. This surge in power demand presents challenges for tech companies committed to powering their data centers with renewable energy sources.

The growing energy demand from AI also raises significant environmental concerns. Google reported a nearly 50% increase in carbon emissions compared to 2019, largely driven by AI-related energy demand. If not managed properly, this could lead to greater reliance on fossil fuels and increased carbon emissions. Additionally, water usage is a concern, as data centers require significant amounts of water for cooling. For example, Google and Microsoft consumed 32 billion liters of water in their data centers in 2022.

While AI's energy consumption continues to grow, efforts to improve efficiency are underway. Improvements in AI software and hardware have led to some gains in efficiency. For instance, Google's data center energy use for machine learning has remained below 15% despite increased AI usage. However, each new generation of AI chips tends to be more energy-intensive, creating a cycle of increasing power demand.

A recent article reports that Microsoft's AI energy demands could lead to the reactivation of a nuclear reactor at Three Mile Island, Pennsylvania. Unit 1, offline for five years, may resume operations by 2028 under an agreement with Constellation Energy. Located near the site of the 1979 partial meltdown, this reactor would supply electricity exclusively to Microsoft for 20 years, pending regulatory approval. Constellation hopes to generate 800 megawatts and add 3,400 jobs. This move highlights nuclear power's potential to meet rising energy demands and support sustainable, large-scale energy solutions.

The Role of Nuclear Power in Meeting AI's Energy Demand

Nuclear power could play a pivotal role in meeting the growing energy demand driven by AI. By integrating nuclear power into the energy mix, we can address the increasing electricity requirements of AI while mitigating environmental impacts. Nuclear power plants can operate continuously, unlike some renewable sources that depend on weather conditions, ensuring a steady supply of electricity for AI systems. This reliability is particularly crucial for data centers, which demand uninterrupted power.

Advancements in nuclear technology, such as small modular reactors (SMRs), offer flexible and scalable solutions that can be deployed more rapidly than traditional large reactors. SMRs can be strategically placed to supply power directly to data centers, reducing transmission losses and improving overall efficiency.

While AI offers tremendous potential for innovation and efficiency, its growing energy appetite presents significant challenges for electricity grids and environmental sustainability. Balancing the benefits of AI with its energy costs will be a crucial task for technologists, energy providers, and policymakers. Nuclear power, with its capacity for large-scale, baseload, low-carbon energy production, stands as a key component in meeting this challenge, ensuring that AI advancements are supported by a sustainable and reliable energy infrastructure.

The Danger of Excluding Nuclear from Our Energy Future

As we advocate for the inclusion of more nuclear power in the global energy mix, we must confront a sobering truth: excluding nuclear from our energy future would be nothing short of disastrous. Without a significant expansion of nuclear power alongside growth in our use of renewables, our addiction to fossil fuels will continue to overheat the planet, pushing us toward the very climate catastrophe we are desperately trying to avoid. The tragic scenarios described at the outset of this book—rising sea levels, extreme weather, major wildfires, and population displacements—would no longer be rare occurrences, but devastating certainties. Human-induced global warming will spiral out of control, with irreversible consequences for both our environment and future generations.

This is not hyperbole. A recent report from the U.S. Energy Information Administration (EIA) presents a clear fork in the road: one path leads to a future where fossil fuels dominate, suffocating our atmosphere with greenhouse gases; the other leads to a clean energy future powered by nuclear alongside renewables. The choice is stark—and the stakes couldn't be higher.

The Catastrophic Cost of Inaction

Turning our backs on nuclear power would be a catastrophic miscalculation, and one that future generations would pay for in profound ways. The cost of inaction is measured not only in economic terms but also in lives lost, ecosystems destroyed, and opportunities squandered. Persisting with fossil fuels will keep greenhouse gas emissions at dangerous, life-threatening levels, making it impossible to meet global climate goals like the Paris Agreement. Inaction would condemn us to a future where environmental degradation accelerates at a breakneck pace, and we lose the ability to reverse course.

Air pollution, driven by the relentless combustion of fossil fuels, already claims millions of lives annually through respiratory diseases, heart conditions, and cancer. But this is just the beginning. As climate change worsens, heatwaves, droughts, and flooding will become even more frequent and severe, causing widespread displacement, food shortages, and escalating conflicts over resources. This isn't some distant possibility—it is the future we are creating right now by failing to act decisively.

But the dangers don't stop there. Renewables like solar and wind, while critical to our energy transition, are inherently intermittent. Without nuclear power to provide steady, reliable baseload energy, our grids will become increasingly unstable. Blackouts and grid failures will become more frequent, particularly during extreme weather events when energy demand spikes. Entire cities could be plunged into darkness, hospitals could lose power, and essential services could be disrupted—putting lives at risk.

Nuclear fuel is extremely dense.

Because of this, the amount of used nuclear fuel is not as big as you think. All of the used nuclear fuel produced by the U.S. nuclear energy industry over the last 60 years could fit on a football field at a depth of less than 10 yards.

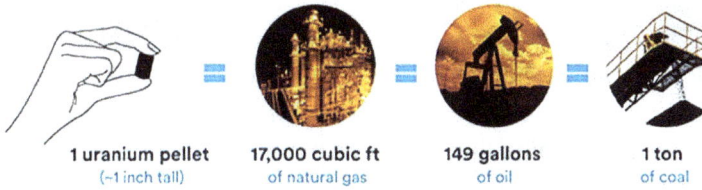

1 uranium pellet	17,000 cubic ft	149 gallons	1 ton
(~1 inch tall)	of natural gas	of oil	of coal

Energy Insecurity: A Growing Threat

By excluding nuclear power, we also leave ourselves vulnerable to energy insecurity. Fossil fuel markets are notoriously volatile, subject to geopolitical tensions, price shocks, and supply chain disruptions. Without the stability of domestically produced nuclear energy, nations will remain at the mercy of these unpredictable forces, risking both economic stability and national security. Countries that fail to diversify their energy portfolios will find themselves increasingly reliant on foreign oil and gas—a dangerous dependency that can undermine sovereignty and expose economies to external shocks.

In contrast, nuclear power offers a secure, long-term solution. It provides a stable, domestic energy source that is immune to the fluctuations of fossil fuel markets. Moreover, nuclear fuel supplies are abundant and geographically diverse, meaning that no single country or region holds a monopoly on the resources needed to power nuclear reactors. This makes nuclear energy not just a tool for fighting climate change but also a strategic asset for ensuring national energy security and resilience.

The Price of Delay

The longer we delay the expansion of nuclear power, the more expensive the consequences become. Every year that we fail to act means more carbon dioxide pumped into the atmosphere, more lives lost to air pollution, and more resources spent on recovering from the growing frequency of climate disasters. The price of delay is steep—and it will only continue to rise.

The time for half-measures has long passed. We cannot rely solely on renewables to power the future. While increasing our solar and wind power is vital, they must be complemented by nuclear energy if we are to achieve the deep decarbonization necessary to avert the worst impacts of climate change. The evidence is clear: we need nuclear energy to provide reliable, carbon-free power at scale, stabilize our grids, and reduce our dependence on fossil fuels. Without it, we are condemning ourselves to a future of instability, insecurity, and irreversible environmental damage.

The Urgency of Action to Stop Climate Change

As we draw this journey to a close, one thing is undeniably clear: nuclear power must play a pivotal role in stopping climate change. This isn't just an option—it's the linchpin in securing a sustainable future for generations to come. If we are truly committed to slashing greenhouse gas emissions and averting a climate catastrophe, harnessing nuclear energy's vast potential isn't just smart—it's essential.

Climate change is no longer a distant threat—it is already a clear and present danger to our planet and the lives of future generations. We are running out of time to curb greenhouse gas emissions, and the need to pivot toward sustainable energy sources has never been more pressing. Among these, nuclear power stands out as the only dependable, large-scale, carbon-free energy option that can decisively alter the course of our future. The clock is ticking, and nuclear power is the solution that can meet the urgency of this moment.

Nuclear Power: The Low-Carbon Dynamo

Nuclear energy isn't just a powerhouse—it is the powerhouse of low-carbon electricity. Unlike fossil fuel plants that pump toxins into our atmosphere, nuclear reactors produce virtually no greenhouse gases during operation. In fact, after more than 18,500 cumulative reactor-years of commercial nuclear power operation across 36 countries, only three major accidents have occurred: Three Mile Island, Chernobyl, and Fukushima. Six decades of experience show that nuclear power is one of the safest methods of generating electricity. The risk of accidents continues to decline with advances in reactor design, and the consequences of an accident or even a terrorist attack are minimal compared to the risks we regularly accept from fossil fuel energy sources.

By scaling up nuclear energy, we can drastically reduce our reliance on fossil fuels, stymie the devastating march of climate change, and provide the energy needed to sustain modern life.

Reliability and Baseload Power: A Critical Edge

Unlike intermittent renewables like solar and wind, which depend on weather conditions, nuclear power delivers a steady, unyielding flow of electricity around the clock. It provides the baseload power our modern grids require, ensuring a stable, resilient energy supply, even during peak demand or adverse weather conditions. This reliability is not a luxury; it's a necessity for a future where clean, consistent energy is paramount. Without it, we cannot achieve the deep decarbonization our planet desperately needs.

Energy Independence and National Security

Nuclear power isn't just about clean energy—it's about security. By diversifying our energy portfolios and reducing dependence on volatile fossil fuel markets, nuclear energy offers a secure, domestic source of power. It shields economies from geopolitical instability and fluctuations in global fuel prices, providing nations with the energy independence required to build a stable and sustainable future.

Technological Advancements: The Future of Nuclear Power

The nuclear sector is advancing rapidly, with innovations that are making nuclear power safer, more efficient, and more sustainable than ever before. Next-generation reactors, such as Generation IV designs and Small Modular Reactors (SMRs), promise greater efficiency, reduced waste, and enhanced safety. These innovations are not hypothetical—they are happening now. With continued investment, these technologies will revolutionize the nuclear industry and redefine what is possible in terms of safe, large-scale energy production.

Global Collaboration: A United Front Against Climate Change

This battle against climate change cannot be won in isolation. It requires a united global effort, one rooted in collaboration and shared innovation. International organizations like the International Atomic Energy Agency (IAEA) and frameworks like the Paris Agreement are essential to fostering the cooperation necessary for scaling up nuclear energy globally. Together, through shared resources and expertise, we can ensure that nuclear power is deployed safely and responsibly, accelerating the path to a decarbonized world.

The NREL Study: A Clear Path to 100% Clean Electricity

The 100% Clean Electricity by 2035 Study (See chapter 3) from the National Renewable Energy Laboratory provides a compelling blueprint for achieving a decarbonized power grid in the United States. This study underscores the urgency of immediate action, showing that rapid deployment of clean energy technologies—including nuclear power—is not only feasible but essential. It highlights how integrating wind, solar, and nuclear energy at an unprecedented scale could provide reliable, low-carbon electricity while unlocking trillions of dollars in societal benefits.

The study is a stark reminder that while the transition to clean energy will require significant investments, the cost of inaction is far greater. By acting now, we can avoid billions in climate-related damages and save thousands of lives by improving air quality. It shows us that a decarbonized future is not just possible—it's within reach. But only if we act with the urgency that this moment demands.

Net-Zero is Not Enough—We Need Nuclear for a Sustainable Future

The climate is heating up faster than anyone could have predicted. We're already feeling the effects of extreme weather events, and the data is clear: time is running out. The clean energy revolution is gaining momentum, but even as we increase our solar, wind, and energy storage technologies, we must recognize a critical truth: achieving net-zero emissions is not the finish line.

Net-zero will certainly help towards stabilizing temperatures, but it won't reverse the committed warming already locked into our atmosphere. And without nuclear power, achieving net-zero will be far more challenging, if not impossible. Moreover, this milestone doesn't address other potent greenhouse gases like methane and nitrous oxide. Nuclear power, with its capacity to provide reliable, large-scale clean energy, is the only solution capable of addressing these challenges head-on.

The ADVANCE Act: A Major Milestone for U.S. Nuclear Energy Policy

On a bright note, the Accelerating Deployment of Versatile, Advanced Nuclear for Clean Energy (ADVANCE) Act, was signed into law in July 2024, representing a significant milestone in U.S. nuclear energy policy. This bipartisan legislation, which has garnered support across political lines, underscores the nation's unified commitment to advancing nuclear energy as a critical tool in the fight against climate change.

Now in its implementation phase, the Act is being actively put into practice by federal agencies, with the Nuclear Regulatory Commission (NRC) taking center stage. The NRC is revising regulations, updating its mission statement, and streamlining licensing processes for new reactors and fuels, all while upholding its core focus on public health and safety. These changes are designed to cut through bureaucratic delays and provide incentives for innovation, such as reduced licensing fees and prize awards, aimed at accelerating growth in the nuclear industry.

In a parallel move, the Department of Energy has launched a $900 million support package to bolster the initial deployment of advanced and small modular reactors, aligning perfectly with the ADVANCE Act's objectives. As the Act's provisions take effect, they are expected to play a transformative role in revitalizing the U.S. nuclear sector, positioning it as a key player in the nation's clean energy future. This legislation could reshape the American energy landscape in the coming years, making nuclear power a cornerstone of the country's carbon reduction strategy.

The Path Forward: Embracing Nuclear Power to Save Our Planet

As we've seen, Nuclear power is not just a complement to renewable energy; it is the essential ally in our fight against climate change. By embracing nuclear energy, we can drastically reduce greenhouse gas emissions, ensure a reliable and low-carbon energy supply, and safeguard the health and security of our planet. With continued technological advancements, shifts in public perception through education, more realistic regulatory approaches, and unwavering international cooperation, we can unlock the full potential of nuclear power—potential that may very well be the key to saving humanity and our planet.

In conclusion, while achieving net-zero emissions is an essential goal, it is only part of the solution. We must adopt a comprehensive approach that addresses all greenhouse gases and tackles the slow, relentless impacts of past emissions. Only then can we fully mitigate the devastating effects of global warming and pave the way for a sustainable future. The time to act is now—nuclear power is the linchpin that can stop climate change. Let's embrace it before it's too late.

A Call to Action: Choosing Nuclear for Our Future

As we stand at this crossroads, the choice before us is unmistakable. Will we continue down the path of fossil fuel dependency, with all the destruction and danger that entails? Or will we embrace nuclear power as an indispensable part of the solution to our climate crisis?

We know the stakes, and we know the solution. Nuclear energy is the only technology that can deliver the reliable, large-scale, low-carbon power we need to meet our climate goals and secure a sustainable future. The *100% Clean Electricity by 2035 Study* has shown us that this future is within reach—but we must act now. The United States' ADVANCE Act is a step in the right direction. However, if we (the global population) fail to include more—much more—nuclear power in our energy future, we are not just making a mistake—we are making a choice. And that choice will determine the fate of our planet.

References

1. *Sam Altman's OpenAI could be entirely powered by nuclear energy in 2027.* (2024, July). Watch. https://www.msn.com/en-us/money/markets/sam-altman-s-openai-could-be-entirely-powered-by-nuclear-energy-in-2027/vi-BB1pSHCx?ocid=socialshare

2. Jones, A. (2024, July 8). *Digital goes green: More data centers migrating to renewable energy.* I.S. Partners. https://www.ispartnersllc.com/blog/data-centers-renewable-energy/

3. *Microsoft's AI uses so much energy that it could bring infamous nuclear plant back into service.* (2024). https://www.msn.com/en-us/news/technology/microsoft-s-ai-uses-so-much-energy-that-it-could-bring-infamous-nuclear-plant-back-into-service/ar-AA1qUAoo?ocid=msedgntp&pc=HCTS&cvid=0c768f908f3340ff89df11f1989c529a&ei=62

4. U.S. Energy Information Administration. (2023). *The ultimate Fast Facts guide to nuclear energy* [Report]. https://www.energy.gov/sites/default/files/2024-02/ne-2023fastfactsguide-021424.pdf

5. *Bill Gates is betting on nuclear fission and fusion to solve the climate crisis.* (2024, September 6). https://www.msn.com/en-us/money/technology/bill-gates-is-betting-on-nuclear-fission-and-fusion-to-solve-the-climate-crisis/ar-AA1q4KTQ?ocid=winp2fpswipe&cvid=78c21485e26f4ae6a06555a4636acc86&ei=7

6. *Outlook for future emissions - U.S. Energy Information Administration (EIA).* (*2023, October 27*). https://www.eia.gov/energyexplained/energy-and-the-environment/outlook-for-future-emissions.php

7. *ADVANCE Act (Accelerating Deployment of Versatile, Advanced Nuclear for Clean Energy Act of 2024).* (2024, September 24). NRC Web. https://www.nrc.gov/about-nrc/governing-laws/advance-act.html

Additional Sources for More Information

Atomic Blender: Looking at the future of nuclear energy, policy, and economics and how they impact our world. Hosted by Michael

Seely, AtomicBlender aims to be the best source for information and perspective on all things related to nuclear. *AtomicBlender*. (n.d.). YouTube. https://www.youtube.com/@atomicblender

Forbes. (2024, October 3). *Nuclear power is finally gaining Favor—But it won't replace fossil fuels anytime soon* [Video]. YouTube. https://www.youtube.com/watch?v=Xg32vUmEITQ

Change Oracle: Since 2008, Change Oracle has provided prescient reviews and insightful analyses of the most important sustainability issues. Change Oracle is widely cited by reports, journals, books, papers, reports, and websites including Forbes, the LA Times, Columbia Law School, Cornell Law Review, and New York Magazine, Change Oracle's transdisciplinary, science-based analyses are read by business leaders, academics, legal scholars, activists, and investors. Change Oracle. (n.d.). *Nuclear power versus renewable energy*. https://changeoracle.com/energy/nuclear-power

BURGES SALMON LLP. (2014). *Glossary of Nuclear Terms* (G. Davies, A. Crathern, & S. Raby, Eds.). Burges Salmon LLP. https://www.nuclearinst.com/write/MediaUploads/Resources/Burges_Salmon_Glossary_of_Nuclear_Terms_-_July_2014_%28FINAL_VERSION%29.pdf

Glossary

Actinides: The group of radioactive metallic chemical elements with atomic numbers from 89 to 103, including thorium, uranium, plutonium, and others. They are used in nuclear reactors, nuclear weapons, and industrial processes.

Advanced Fuel Cycles: Alternative methods for managing and recycling nuclear fuel to improve efficiency, reduce waste, and enhance safety in nuclear power generation.

Biomass: Organic matter, such as plants or plant-based materials, used as a renewable energy source for generating heat, electricity, or biofuels.

Breeder Reactor: A specialized type of nuclear reactor designed to produce more fissionable material than it consumes while generating energy. Breeder reactors convert fertile isotopes like uranium-238 into fissile plutonium-239, significantly extending the nuclear fuel supply beyond conventional reactors.

Climate Change: Refers to long-term shifts in temperature and weather patterns, primarily caused by human activities like burning fossil fuels, but also includes natural variability. Climate change encompasses global warming, sea-level rise, and extreme weather effects.

Closed Nuclear Cycle: A nuclear fuel cycle in which spent nuclear fuel is reprocessed to extract fissile materials, such as uranium and plutonium, for reuse in reactors. This approach reduces the amount of high-level radioactive waste and maximizes the efficiency of the fuel supply by recycling valuable isotopes, as opposed to the "open" or "once-through" cycle, where spent fuel is disposed of without reprocessing.

Control Rods: Devices made of neutron-absorbing materials, such as boron or cadmium, used to control the rate of fission reactions in a nuclear reactor.

Convection: In nuclear reactor safety, convection is the process by which heat is transferred through the movement of fluids (such as water or gas) within a reactor system. In nuclear reactors, convection plays a critical role in cooling the reactor core by circulating coolant, preventing overheating, and maintaining safe operating temperatures. Passive safety systems often rely on natural convection, where coolant circulates without the need for mechanical pumps, enhancing reactor safety in case of system failure.

Coolant: A substance, typically a liquid or gas, used to remove heat from a system, such as a nuclear reactor, to maintain its operating temperature.

Criticality: The condition where a nuclear reactor's chain reaction is self-sustaining, with the neutron population remaining steady. Achieving and maintaining criticality is essential for the continuous production of energy.

Depleted Uranium (DU): A byproduct of uranium enrichment, primarily composed of uranium-238 with lower levels of uranium-235. It is about 40% less radioactive than natural uranium and is extremely dense, making it useful for military applications like armor-piercing ammunition and in some industrial uses.

Fast Neutrons: High-energy neutrons with kinetic energies typically above 1 MeV, produced through nuclear fission, fusion, and other nuclear processes. They play a crucial role in fast-neutron reactors and breeding reactions, offering advantages in efficiency and waste reduction.

Fast Neutron Spectrum: Fast neutron reactors, such as sodium-cooled fast reactors, operate with a fast neutron spectrum, meaning neutrons with higher energies compared to thermal neutrons used in conventional light-water reactors. This allows for more efficient fission of fertile isotopes like uranium-238 and the breeding of fissile plutonium-239 from uranium-238.

Fissile: In nuclear physics, fissile refers to materials that can undergo fission with low-energy thermal neutrons, such as uranium-235, uranium-233, and plutonium-239. This property is essential for sustaining nuclear chain reactions in reactors and nuclear weapons.

Fission: A nuclear reaction in which the nucleus of an atom splits into two or more smaller nuclei, releasing a large amount of energy in the process.

Footprint: In renewable energy projects, footprint refers to the physical land area occupied or impacted by energy generation infrastructure.

Fusion: A nuclear reaction where two light atomic nuclei combine to form a heavier nucleus, releasing significant energy. It powers stars and has potential as a clean energy source with minimal radioactive waste.

Gas Centrifugation: A method used to enrich uranium by separating isotopes using the difference in mass through centrifugal force.

Gaseous Diffusion: A method used to enrich uranium by separating isotopes based on their different diffusion rates through a porous barrier.

Greenhouse Gases: Gases that trap heat in the atmosphere, warming the Earth. Key gases include carbon dioxide, methane, and nitrous oxide. Human activities have increased their levels, causing global warming and climate change.

Half-Life: The amount of time it takes for half of a radioactive material to decay into a stable form is known as its half-life. For instance, the half-life of plutonium-239, a prevalent isotope in high-level waste, is roughly 24,000 years. While the radioactivity decreases over time, the material must be safely stored until it decays to levels considered safe.

Heavy Water (Deuterium Oxide): Water in which the hydrogen atoms are replaced by deuterium, a heavy isotope of hydrogen. It is used as a neutron moderator in certain types of nuclear reactors.

High-Level Radioactive Waste (HLW): Highly radioactive waste material from nuclear reactors, including spent fuel. It requires secure containment for thousands of years due to its long-lived radioactivity.

Isotopes: Different forms of the same chemical element that have the same number of protons but different numbers of neutrons in their atomic nuclei. While isotopes of the same element share chemical properties, they can have different physical properties and levels of stability, with some being radioactive.

Levelized Cost of Electricity (LCOE): The commonly used metric for factoring in the total costs of building, operating, and maintaining a power plant over its expected lifetime, as well as the amount of electricity produced.

Low-Level Radioactive Waste (LLW): Waste materials like clothing and tools contaminated with low levels of radioactivity. They require less stringent disposal methods and typically decay to safe levels over time.

Megawatt (MW): A unit of power equal to one million watts. It measures the rate at which energy is produced or consumed at any given moment. MW is commonly used to express the capacity or output of power plants, regardless of the energy source, such as coal, hydroelectric, solar, or nuclear.

Megawatt Electrical (MWe): A megawatt electrical refers specifically to the electrical power output of a power plant. MWe indicates the actual electrical energy available for use after accounting for inefficiencies and losses in converting thermal energy to electrical energy.

Meltdown: A nuclear meltdown is a severe accident where a reactor's core overheats and melts due to inadequate cooling. This can damage fuel rods, release radioactive materials, and potentially breach containment. Modern reactors have safety systems to prevent or mitigate such events.

Metric Ton: A unit of mass equal to 1,000 kilograms (approximately 2,204.62 pounds). It is commonly used in global industry and science for measuring quantities of materials, including nuclear fuel and radioactive waste.

Moderator: A substance, such as water or graphite, used in a nuclear reactor to slow down fast neutrons and increase the probability of nuclear fission.

MOX Fuel: Mixed Oxide nuclear fuel contains a blend of oxides of plutonium and uranium and is an alternative to the low-enriched uranium fuel commonly used in light-water reactors. It typically consists of plutonium dioxide mixed with either natural uranium, reprocessed uranium, or depleted uranium dioxide.

Neutron: A subatomic particle found in the nucleus of an atom. Neutrons have no electric charge and help stabilize the nucleus, balancing the repulsive forces between positively charged protons.

Nuclear Weapon States (NWS): Countries defined under the Nuclear Nonproliferation Treaty as states that had manufactured and detonated a nuclear explosive device prior to January 1, 1967.

Nuclei: Plural form of nucleus.

Nucleus: The central part of an atom that contains protons and neutrons.

Photovoltaic (PV): A technology used for solar power, referring to the direct conversion of sunlight into electricity using semiconductor materials that exhibit the photovoltaic effect.

Plasma: In nuclear fusion, plasma is an extremely hot, ionized gas consisting of positively charged ions and free electrons. It provides the essential environment for fusion reactions, allowing atomic nuclei to overcome repulsion and fuse. Plasma is naturally found in stars and is crucial for developing controlled fusion on Earth as a potential energy source.

Plutonium: A radioactive chemical element with the symbol Pu and atomic number 94. It is primarily used in nuclear weapons and as fuel for certain types of nuclear reactors.

Radioactive: The property of certain materials to emit radiation resulting from the decay of unstable atomic nuclei.

Reactivity: The measure of how far a nuclear reactor is from its critical state. Positive reactivity indicates an increase in power output, while negative reactivity indicates a decrease. Managing reactivity is crucial for safe reactor operations.

Spent Nuclear Fuel: Fuel that has been used in a nuclear reactor and is no longer capable of sustaining a chain reaction. It is highly radioactive and requires proper disposal or reprocessing.

Thorium: A naturally occurring radioactive element with the symbol Th and atomic number 90. It is an alternative fuel source for nuclear reactors and has potential as a future energy source.

Tokamak: A device using magnetic fields to confine plasma in a ring shape for controlled nuclear fusion power.

Tritium: A radioactive isotope of hydrogen with one proton and two neutrons. With a half-life of 12.32 years, tritium emits beta radiation. It is used in nuclear weapons, fusion research, self-powered lighting, and as a tracer in scientific studies. It can form tritiated water and plays a key role in certain fusion reactions.

Uranium: A naturally occurring heavy metal with the symbol U and atomic number 92. It is primarily used as a fuel in nuclear reactors and nuclear weapons due to its radioactive properties.

Warheads: Nuclear warheads are explosive devices at the core of nuclear weapons, designed to produce nuclear explosions. They come in fission and fusion types, containing fissile material and conventional explosives. Warheads can be deployed on missiles or aircraft, with their power measured in kilotons or megatons.

About the Author

Mary Fran Reed PhD, Scientist & World Sailor

Dr. Reed is an expert in nuclear sciences and emergency planning, with a distinguished career spanning research, teaching, and public service. Her journey into the world of nuclear science began at the Lawrence Radiation Laboratory, where her father's work as a technical writer ignited her interest in the field. As a high school student, she spent several summers working as a radiochemistry technician at the same laboratory, which ignited her lifelong passion for nuclear science and the peaceful uses of atomic energy.

Reed pursued her academic ambitions at the University of California, Berkeley, where she earned her PhD in nuclear chemistry, specializing in nuclear reaction mechanisms. Her research contributed to the understanding of nuclear processes.

Following her doctorate, Reed began her teaching career at the University of Kentucky, where she instructed students in chemistry before earning certification as a Nuclear Medicine Physicist. She went on to teach medical residents about the physics of nuclear medicine and became an active researcher, publishing and presenting work in the field.

Her career took a critical turn when she returned to California to lead the state's nuclear power plant emergency response planning program and to coordinate efforts between plant operators, local governments, and federal agencies to ensure public safety through meticulous planning and response strategies.

Parallel to her professional achievements, Reed cultivated a love for sailing, which later led to an extraordinary global circumnavigation with her husband. Together, they visited 42 countries, documenting their adventures and experiences, which inspired her book, *DREAMING OF OCEAN CRUISING? Sailing off into the Sunset? What You Ought to Know!*

In her latest work, *ATOMIC GREEN: Nuclear Power Can Stop Climate Change,* Reed draws on decades of expertise to advocate for nuclear energy as an essential solution to the climate crisis. She dispels myths surrounding nuclear power and presents a compelling case for its role in reducing greenhouse gas emissions, emphasizing modern advancements in safety and reactor design. Reed's writing educates and empowers readers, making a powerful argument for nuclear energy's role in stopping climate change and securing a sustainable future.